Understanding Vertebrate Zoology

Understanding Vertebrate Zoology

Ezra Samberg

Larsen & Keller
www.larsen-keller.com

Understanding Vertebrate Zoology
Ezra Samberg
ISBN: 978-1-64172-706-8 (Hardback)

Larsen & Keller
Published by Larsen and Keller Education,
5 Penn Plaza,
19th Floor,
New York, NY 10001, USA

Cataloging-in-Publication Data

Understanding vertebrate zoology / Ezra Samberg.
p. cm.
Includes bibliographical references and index.
ISBN 978-1-64172-706-8
1. Vertebrates. 2. Zoology. 3. Animals. I. Samberg, Ezra.
QL605 .U53 2021
596--dc23

For more information regarding Larsen and Keller Education and its products, please visit the publisher's website www.larsen-keller.com

Table of Contents

Preface

The animals which have a backbone are known as vertebrates. The biological discipline which consists of the study of vertebrate animals is known as vertebrate zoology. There are various branches within this discipline such as mammalogy, ornithology, ichthyology, herpetology and batrachology. Mammology is a field of study which focuses on mammals. Ornithology deals with the study of birds. Different reptiles are studied under the discipline of herpetology and ichthyology is concerned with the study of fish. There are various subdisciplines within these fields with different specializations. The topics included in this book on vertebrate zoology are of utmost significance and bound to provide incredible insights to readers. Some of the diverse topics covered herein address the varied branches that fall under this category. The textbook is appropriate for students seeking detailed information in this area as well as for experts.

To facilitate a deeper understanding of the contents of this book a short introduction of every chapter is written below:

Chapter 1- The species of animals which fall under the subphylum vertebrata are called vertebrates. They are all chordates who possess a backbone. The branch of biology which studies vertebrate animals is known as vertebrate zoology. This chapter has been carefully written to provide an easy understanding of the varied facets of vertebrates and vertebrate zoology.

Chapter 2- The discipline of zoology which focuses on the study of animals which belong to the class Pisces, commonly known as fish, is termed as ichthyology. The different categories of fish which are studied within this field include agnathan, chondrichthyes and osteichthyes. This chapter discusses in detail these types of fish to provide an extensive understanding of the subject.

Chapter 3- The vertebrate animals which belong to the class Mammalia are called mammals. The study of mammals is called mammalogy. Some of the different classifications within mammals are Theria and Prototheria. The topics elaborated in this chapter will help in gaining a better perspective about these types of mammals.

Chapter 4- The branch of zoology which focuses on the study of animals which belong to the class Aves, also called birds, is termed as ornithology. Birds are broadly divided into two groups, ratites and non-ratites. The chapter closely examines the key aspects of these types of birds to provide an extensive understanding of the subject.

Chapter 5- The domain of biology which studies the amphibians and reptiles is known as herpetology. Batrachology is a sub-discipline of herpetology which focuses on amphibians such as salamanders and frogs. The topics elaborated in this chapter will help in gaining a better perspective about these animals studied under herpetology and batrachology.

I would like to share the credit of this book with my editorial team who worked tirelessly on this book. I owe the completion of this book to the never-ending support of my family, who supported me throughout the project.

Ezra Samberg

1

Introduction to Vertebrate Zoology

The species of animals which fall under the subphylum vertebrata are called vertebrates. They are all chordates who possess a backbone. The branch of biology which studies vertebrate animals is known as vertebrate zoology. This chapter has been carefully written to provide an easy understanding of the varied facets of vertebrates and vertebrate zoology.

Vertebrates

Vertebrate, also called Craniata is any animal of the subphylum Vertebrata, the predominant subphylum of the phylum Chordata. They have backbones, from which they derive their name. The vertebrates are also characterized by a muscular system consisting primarily of bilaterally paired masses and a central nervous system partly enclosed within the backbone.

The subphylum is one of the best known of all groups of animals. Its members include the classes Agnatha, Chondrichthyes, and Osteichthyes (all fishes); Amphibia (amphibians); Reptilia (reptiles); Aves (birds); and Mammalia (mammals).

General Features

Although the vertebral column is perhaps the most obvious vertebrate feature, it was not present in the first vertebrates, which probably had only a notochord. The vertebrate has a distinct head, with a differentiated tubular brain and three pairs of sense organs (nasal, optic, and otic). The body is divided into trunk and tail regions. The presence of pharyngeal slits with gills indicates a relatively high metabolic rate. A well-developed notochord enclosed in perichordal connective tissue, with a tubular spinal cord in a connective tissue canal above it, is flanked by a number of segmented muscle masses. A sensory ganglion develops on the dorsal root of the spinal nerve, and segmental autonomic ganglia grow below the notochord. The trunk region is filled with a large, bilateral body cavity (coelom) with contained viscera, and this coelom extends anteriorly into the visceral arches. A digestive system consists of an esophagus extending from the pharynx to the stomach and a gut from the stomach to the anus. A distinct heart, anteroventral to the liver, is enclosed in a pericardial sac. A basic pattern of closed circulatory vessels is largely preserved in most living forms. Unique, bilateral kidneys lie retroperitoneally (dorsal to the main body cavity) and serve blood maintenance and excretory functions. Reproductive organs are formed from tissue adjacent to the kidneys; this original close association is attested by the tubular connections seen in males of living forms. The ducts of the excretory organs open through the body wall into a cloacal chamber, as does the anus of the digestive tract. Reproductive cells are shed through nearby abdominal pores or through special ducts. A muscular tail continues the axial musculature of the trunk.

Approximately 45,000 living species constitute the vertebrates. Species of several classes are found from the high Arctic or Antarctic to the tropics around the Earth; they are missing only from interior Antarctica and Greenland and from the North Polar ice pack. In size, vertebrates range from minute fishes to elephants and whales (of up to 100 tons), the largest animals ever to have existed. Vertebrates are adapted to life underground, on the surface, and in the air. They feed upon plants, invertebrate animals, and one another. Vertebrate faunas are important to humans for food and recreation.

In order to give a broad and comparative view of their life histories, the vertebrates are subdivided here into major groups based on morphology: the cyclostomes (jawless fishes), the chondrichthyes (cartilaginous fishes), the teleostomes (bony fishes), and the tetrapods.

Cyclostomes

The cyclostomes include two classes of living, jawless fishes (agnathous)—Petromyzontiformes (lamprey eels) and Myxiniformes (hagfishes). The hagfishes are totally marine, often living in deep waters associated with muddy bottoms. The lampreys may be marine as adults but spawn in fresh waters, where the larvae spend some time before metamorphosing to the adult. Some lampreys live entirely in fresh water and may change only slightly in habit as a result of metamorphosis. Without lateral fins, lampreys swim by undulations of the body and can control direction only for short distances.

Lamprey: Lamprey (Lampetra) on rainbow trout.

The living agnaths are predatory, the lampreys being well known for attacking salmonoid fishes. The lamprey attaches to its prey using its round, suctorial mouth, and it rasps a hole through the outer tissues using a tongue armed with keratinized teeth. It suctions off bits of tissue, blood, and body fluids. The hagfishes feed somewhat similarly, but on a variety of prey—invertebrates (worms and soft-bodied forms) and dead fishes.

The lampreys produce small eggs, which develop directly into larvae that burrow into the muddy bottom of the stream. With its mouth at the surface of the mud, the larva filter feeds until large enough to metamorphose and swim off as a small adult. In contrast, the hagfishes produce relatively large encapsulated, yolky eggs up to two centimetres in length. When laid, these eggs attach to any available object by terminal hooks. The encased egg develops more or less directly into a miniature adult.

Chondrichthyes

The sharks, rays, and chimaerids are usually marine, but some sharks have entered fresh waters (the Amazon) or even live there permanently (Lake Nicaragua). In size, sharks range from the whale shark, nearly 10 metres in length, to rather small species, three centimetres in length. They usually weigh 25 to 200 kilograms (55 to 440 pounds). Sharks are predatory animals. Some large shark species (basking and whale sharks) filter feed on small crustaceans. Herbivorous sharks are unknown. Sharks swim by undulations of the tail, but rays "fly" through the water by undulations of the pectoral fins. Most species occur in near-shore waters, but some range widely throughout the oceans. A few are found in deep water.

Southern stingrays (Dasyatis americana).

A few sharks produce live young (viviparous) after internal fertilization. The posterior angle of the male's pelvic fins are modified into a clasper, which acts as an intromittent organ in copulating with the female. Most sharks lay large yolky, encapsulated eggs with hooks for attachment. The young develop directly and begin life as miniature adults. The young that develop in the mother's uterus obtain nutrients from the large yolk sac until they are born alive. In a few cases, the uterine wall secretes nutrients.

The teleostome, or osteichthyian, fishes (those having an internal bony skeleton) can be divided into two groups: the subclasses Actinopterygii (ray-finned fishes) and Sarcopterygii (lobe-finned fishes). The latter group includes the lungfishes, which live in marshes, ponds, or streams, and are frequent air breathers. They lay fairly large eggs, with a limited amount of yolk, that are enclosed in jelly coats like those of an amphibian. The eggs develop into small fishes that feed on live prey. The larvae of the African lungfish have external gills to supplement oxygen intake.

Teleostomes

Actinopterygian fishes are the common bony fishes of modern aquatic environments. They range in size from fishes that are only millimetres in size to those two or more metres (6.6 or more feet) in length, weighing 500 kilograms or more. Large species (sturgeons) are found in fresh waters (several other large species are found in the Amazon) as well as in marine environments. The diet may include plants, animals, and carrion. Most species are midwater swimmers, but many spend much time lying on the bottom. Tail, pectoral, and even dorsal fins are used in swimming. Reproduction in this group is by way of large numbers of small eggs, which produce small larvae or develop directly to the adult.

American paddlefish (Polyodon spathula).

Tetrapods

The tetrapods live primarily on land and are rather similar in habit. Members include the amphibians, reptiles, birds, and mammals. Amphibians are widespread in the warmer parts of the continents, being absent only in the far north and in the Antarctic. Three orders are recognized: Candata (the salamanders), the frogs and toads (Anura, or Salientia), and the Apoda or Gymnophiona (caecilians). Modification takes many forms, from the moist glandular skin (some scale remnants persist in apodans) to the loss of many of the bones of the skull. Like their ancestors, amphibians are cold-blooded and tend to be aquatic or limited to moist surroundings. Salamanders are seemingly the least modified in body form. They do not actively pursue prey and at best are only marginal swimmers. In swimming or crawling, the salamander's body and tail undulate. Frogs and toads hop using hind-limb propulsion and the forelimbs as body props. This dominance of the hind limb in locomotion is best seen in swimming when the forelimbs are drawn back against the body. In contrast to the salamanders and frogs, the burrowing, wormlike apodans are without limbs.

European pond turtle (Emys orbicularis).

Amphibians usually trap food using a tongue that can be shot out of the mouth, or they use the mouth itself to grasp and ingest food. There is great variation in foods; only the larvae of frogs and toads appear to be plant feeders, a specialization that is reflected in the highly modified jaws and guts of the tadpoles.

Amphibians have retained a simple egg cell with a gelatinous cover. The eggs are laid in ponds, streams, or even in damp places high in trees, usually in great numbers. Fertilized eggs develop into free-swimming larvae, which then metamorphose to adults, but in highly specialized forms.

The class Reptilia retains many of the structural characteristics of the ancestral amphibian. While most reptiles are carnivorous, feeding on other organisms, a few are herbivorous (e.g., tortoises). As cold-blooded animals, reptiles tend to be limited to temperate and tropical areas, but, where found, they are relatively common, although not as large or conspicuous as birds or mammals. Most reptiles are terrestrial, but a few are aquatic. As basic tetrapods, reptiles move about by creeping or swimming in a fashion similar to amphibians. Some reptiles, however, can lift the body from the ground and run rapidly either in a quadrupedal or bipedal fashion. Reptiles lay relatively large, shelled eggs. In a few instances, the eggs and young are cared for by the female; in others, the young are born alive (ovovivipary).

Birds are warm-blooded, and, although most are capable of flight, others are sedentary and some are flightless. Like their relatives the reptiles, birds lay shelled eggs that differ largely in the amount of calcification (hardening) of the shell. The young are usually cared for in a nest until they are capable of flight and self-feeding, but some birds hatch in a well-developed state that allows them to begin feeding immediately or even take flight. The megapods lay their eggs in mounds of rotting vegetation, which supplies the heat for incubation. (Nesting activities similar to those of some birds are seen in the crocodilians).

The mammals range in size from tiny shrews or small bats weighing only a few grams to the largest known animals, the whales. Most mammals are terrestrial, feeding on both animal and vegetable matter, but a few are partially aquatic or entirely so, as in the case of the whales or porpoises. Mammals move about in a great variety of ways: burrowing, bipedal or tetrapedal running, flying, or swimming. Reproduction in mammals is usually viviparous, the young developing in the uterus, where nutritive materials are made available through an allantoic placenta or, in a few cases, a yolk sac. The fertilized egg develops directly into the adult. The monotremes (platypus and echidna) differ from other mammals in that they lay eggs which hatch, and the relatively undeveloped young are carried in a pouch or kept in a nest; the growing young lap up a milk nutrient fluid exuded from the belly of the mother.

Form and Function

External Features

The evolution of the notochord, dorsal nerve tube, and pharyngeal slits in chordate structure suggests improved swimming capability and probably greater ability to capture prey. Specialization in the vertebrate for the active capture of larger prey is evident both in the structure of the mouth and in the relatively simple structure of the pharynx, with its strong gill development. Specialization for feeding is again seen in the two basic groups of vertebrates, the agnathans and gnathostomes. Swimming adaptations are also numerous and involve variations both in body form and in medial fins and the two pairs of lateral fins.

Internal Features

Skeletal System

Support and protection are provided by the exoskeletal and endoskeletal divisions of the skeletal system. The exoskeleton, when present, is basically protective but functions in tooth support in the mouth region. The endoskeleton protects the brain and spinal cord and assists primarily with locomotion in the trunk and tail regions. The endoskeleton begins as cartilage and may remain so or may develop into bone. The cartilaginous endoskeleton, found in the shark or chimaerid, is usually calcified so as to be stiffer and stronger. Bone is distinctive but highly variable; some types of bone contain cells, others do not, or the bone may be laminar, spongy, or arranged in sheathing layers around blood channels.

Tissues and Muscles

Tissue development in the vertebrate is unique in its complexity; tissues in the strict sense (defined as a mass or sheet of similar cells with a similar function), however, do not exist. The simplest situation is seen in the epidermis, but even here there is a layered system in which different cell types provide different functions (such as protection and secretion). The stratified epithelium of the vertebrate is highly characteristic of that group (a similar one is seen in only one invertebrate group, the class Chaetognatha).

Other tissues of the vertebrate are more complex than the epithelium. For example, skeletal muscle consists not only of striated muscle fibres but also of connective tissue, which binds it together and attaches it by way of tendons. This contractible tissue includes nerves and blood vessels and their contained blood. Skeletal muscles thus appear as simple organs, just as do the smooth muscles in the wall of the gut or the iris muscles of the eye. Such unique histological complexity runs through the entire body of the vertebrate.

Nervous System and Organs of Sensation

The dorsal position, tubular structure, and epidermal origin of the central nervous system are definitive of the chordates, although some may see similarities with the hemichordates. The sensory structures are distinctive of the chordates and include the paired nasal, optic, and otic organs (along with the strongly differentiated head).

The nasal vesicle is variously open to the environment, and its sensory cells, as chemical receptors, are not unlike those in the taste buds of the mouth. The eye is the most complex organ of the head and is a lateral outpocketing of the anterior end of the brain tube. Later it acquires a lens of epidermal origin. The act of focusing the eye (accommodation) shows extensive adaptive variation among the different groups of vertebrates.

The otic vesicle starts from a simple sac formed by the invagination of an ectodermal placode. These developmental changes also include the changes of innervation. Whereas the original structure was basically an equilibrium adaptation, other functions, such as an awareness of movement or the sensation of the proximity of prey, developed.

The lateral-line system of canals and sensory organs is a unique vertebrate feature. The elements of this system are found on the head as well as the body. This system is related to the ear and presumably at its origin served a similar function. This system is lost in terrestrial vertebrate forms.

Digestive System

The digestive system of the vertebrate is distinctive in its structure but not in its function. The mouth and pharynx can be considered as parts of this system; the latter as an expanded cavity in the head is unmatched in any other group.

Presumably the original condition of the digestive glands was that of a ventral diverticulum which may have received the food mass into its cavity. This diverticulum, matched by the diverticulum seen in the amphioxus or the "intestine" of the tunicate, produced the secretions (bilelike and enzymes) of both liver and pancreas. Through time, the liver gradually differentiated from the pancreas. The size and separation of the liver from the gut suggest its separate blood and metabolic activities. The most obvious by-product of the liver, bile, necessitated the formation of a gall bladder and a duct connection with the gut. The pancreas, in contrast, continued to produce digestive enzymes, but its secretory cells were no longer in direct contact with the food mass. Because the pancreas was only a partial source of intestinal digestive enzymes, it was sometimes reduced in size and enclosed in the gut wall itself (agnaths) or dispersed as tiny bits of tissue in the mesentary supporting the gut (actinopterygians).

Excretory System

The excretory system is unique in its nephrons, which filter the blood in the glomeruli and remove a variety of wastes from the body through selective secretion and reabsorption. In the shark or the coelacanth Latimeria, urea is used to raise the osmotic pressure of the blood to that of the marine habitat, thus saving these organisms considerable metabolic energy. The large intestine (sometimes centred in a rectal gland) acts as an auxiliary excretory organ, as do also the gills of fishes or the sweat glands of mammals.

Respiration and Gas Exchange

Respiration, like excretion, involves specialized body structures, such as lungs or gills, but also can involve other areas, such as the skin itself. Respiration involves exchange of gases both between the body of the organism and the environment and between the blood system and the body tissues. It also involves cellular respiration where oxygen is used and carbon dioxide is produced. There is nothing characteristic of the vertebrate in this functional area; even the hemoglobin of the blood is suggested in the respiratory pigments of other animals.

Circulatory System

circulatory system of vertebrates is closed in that fluids course through vessels, but there is free movement of cells in and out of blood. Some leukocyte (white blood cell) movement out of the capillaries and fluid leakage are observed in all tissues. Blood tissues are distinctive in the range of specialized cells, although these vary in detail among animals. The immune function of the blood is best developed in the vertebrate.

The Endocrine System

The endocrine system is characterized by its separate organs. The occurrence of a pituitary or a thyroid gland is suggestive of the evolutionary change and specialization that took place within this group. The relatively unspecialized nature of some parts of this system is seen in certain scattered cells in the gut wall or even the clumps of islet cells of the pancreas.

Evolution and Paleontology

The knowledge of vertebrates as revealed by fossils has grown rapidly during the past few decades, but there is much still to be discovered. The ancestral vertebrate (protovertebrate) has been sought for more than 100 years, and the likelihood of finding it today is not much greater than in the past. It can be assumed that the protovertebrate was small and soft-bodied, two factors that suggest the improbability of finding a fossilized form in a recognizable condition. There are Cambrian fossils that have been suggested to be fossil cephalochordates and there are scales of agnath fishes, but the first type of fossil is too simple and the second already too complex to explain the transition.

Vertebrate Zoology

Vertebrate Zoology is the discipline of biology which focuses on then study of animals that fall in the phylum vertebrata. These animals have backbone in their body. Examples of such animals include fish, amphibians, reptiles, birds and mammals. It have many subdivisions which includes Ichthyology, Mammalogy, Ornithology, Herpetology, Batrachology, and these divisions are further subdivisions.

Chordates (Chordata)

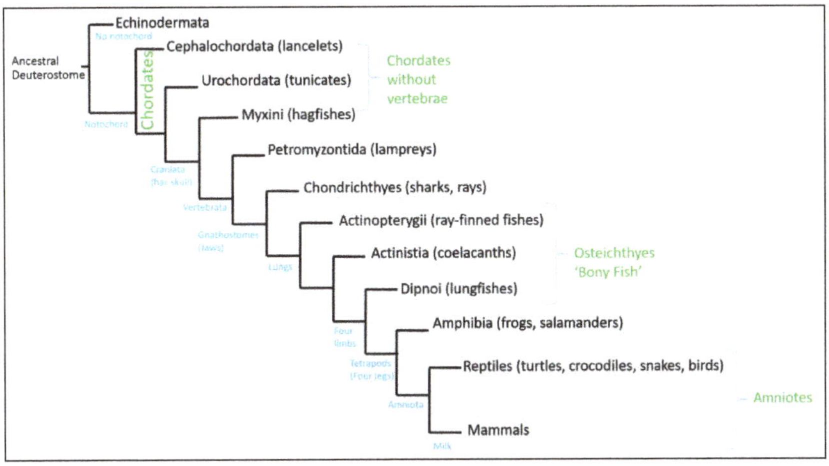

Overarching phylogenetic tree of chordates and their ancestor. Note that this is one hypothesis for the evolution of animals based on the criteria shown in blue. Common group names referring to several phyla are shown in green.

Animals in the phylum Chordata share four key features that appear at some stage during their development:

1. A notochord,
2. A dorsal hollow nerve cord,

3. Pharyngeal slits, and
4. A post-anal tail.

The chordates are named for the notochord, which is a flexible, rod-shaped structure that is found in the embryonic stage of all chordates and in the adult stage of some chordate species. It is located between the digestive tube and the nerve cord, and provides support through the length of the body.

The dorsal hollow nerve cord derives from ectoderm that rolls into a hollow tube during development. In chordates, it is located dorsal to the notochord. In contrast, other animal phyla are characterized by solid nerve cords that are located either ventrally or laterally. The nerve cord found in most chordate embryos develops into the brain and spinal cord, which compose the central nervous system.

Pharyngeal slits are openings in the pharynx (the region just posterior to the mouth) that extend to the outside environment. In organisms that live in aquatic environments, pharyngeal slits allow for the exit of water that enters the mouth during feeding. Some invertebrate chordates use the pharyngeal slits to filter food out of the water that enters the mouth. In vertebrate fishes, the pharyngeal slits are modified into gill supports, and in jawed fishes, into jaw supports. In tetrapods (amphibians, reptiles, birds, and mammals), the slits are modified into components of the ear and tonsils. Tetrapod literally means 'four-footed,' which refers to the phylogenetic history of various groups that evolved accordingly, even though some now possess fewer than two pairs of walking appendages.

The post-anal tail is a posterior elongation of the body, extending beyond the anus. The tail contains skeletal elements and muscles, which provide a source of locomotion in aquatic species, such as fishes. In some terrestrial vertebrates, the tail also helps with balance, courting, and signaling when danger is near. In humans, the post-anal tail is vestigial, that is, reduced in size and nonfunctional.

Non-vertebrate chordates: Cephalochordata (lancelets), Urochordata (Tunicates), and Myxini (hagfishes).

These groups are the chordates which do not possess vertebrae. Many are hermaphroditic, sessile or buried within the sand of aquatic environments, and hatch from eggs within the parent's body. Some undergo metamorphosis into their adult form. Most are suspension-feeders, feeding on algae or small invertebrates.

In particular, the first fish, Myxini (hagfish) fall within this group. They possess a 'skull', that is, a bony, cartilaginous, or fibrous structure encasing the head. However, they lack a backbone, the key feature defining vertebrates.

Vertebrate Chordates

Vertebrates display the four characteristic features of the chordates; however, members of this group also share derived characteristics that distinguish them from invertebrate chordates. Vertebrata is named for the vertebral column, composed of vertebrae, a series of separate bones joined together as a backbone. In adult vertebrates, the vertebral column replaces the notochord, which is only seen in the embryonic stage.

Vertebrates are the largest group of chordates, with more than 62,000 living species categorized based on anatomical and physiological traits. Here we will consider the traditional groups which constitute classes in the subphylum Vertebrata. Note that many modern authors classify birds within Reptilia, which correctly reflects their evolutionary heritage. We consider them separately only for convenience.

Vertebrate Groups

Fishes were the earliest vertebrates, with jawless species being the earliest and jawed species evolving later. They are active feeders, rather than sessile, suspension feeders.

Agnathostomes (Jawless Fishes)

Jawless fishes, such as lampreys, represent the first true vertebrate lineage. They are craniates that represent an ancient vertebrate lineage that arose over one half-billion years ago. A defining feature is the lack of paired lateral appendages (fins), such that they appear tubular.

Gnathostomes (Jawed Fishes)

Gnathostomes or 'jaw-mouths' are vertebrates that possess jaws. One of the most significant developments in early vertebrate evolution was the development of the jaw, which is a hinged structure attached to the cranium that allows an animal to grasp and tear its food. Early gnathostomes also possessed two sets of paired fins, allowing the fishes to maneuver accurately and become mobile predators. These two traits allowed early gnathostomes to exploit food resources that were unavailable to jawless fishes. Most modern fishes are gnathostomes.

Dunkleosteous was an enormous placoderm from the Devonian period.
It measured up to 10 meters in length and weighed up to 3.6 tons.

Chondrichthyes (Cartilaginous Fishes)

The clade Chondrichthyes is diverse, consisting of sharks, rays, and skates, together with sawfishes and a few dozen species of fishes called chimaeras (ghost sharks). Chondrichthyes are jawed fishes that possess paired fins and a skeleton made of cartilage. They are thought to be descended from the placoderms, which had skeletons made of bone; thus, the cartilaginous skeleton of Chondrichthyes is a later development.

Most cartilaginous fishes live in marine habitats, with a few species living in fresh water for a part or all of their lives. Most sharks are carnivores that feed on live prey, either swallowing it whole or using their jaws and teeth to tear it into smaller pieces, but some are suspension feeders that feed on plankton.

Hammerhead sharks tend to school during the day and hunt prey at night.

Osteichthyes (Bony Fishes)

The vast majority of present-day fishes belong to this group, which consists of approximately 30,000 species, making it the largest class of vertebrates in existence today. The 'bony fish' group includes the Actinopterygii (ray-finned fishes), Actinistia (coelacanths), and Dipnoi (lungfish).

Nearly all bony fishes have an ossified skeleton. This characteristic has only reversed in a few groups of Osteichthyes, such as sturgeons and paddlefish, which have primarily cartilaginous skeletons. The skin of bony fishes is often covered by overlapping scales, and glands in the skin secrete mucus that reduces drag when swimming and aids the fish in osmoregulation.

Amphibians

Amphibians are vertebrate tetrapods that include frogs, salamanders, and caecilians. Most species must remain forever associated with moist environment, at a minimum in early development as eggs.

As tetrapods, most amphibians are characterized by four well-developed limbs, although a few clades have since lost some or all limbs. An important characteristic of extant amphibians is a moist, permeable skin that is achieved via mucus glands that keep the skin moist; thus, exchange of oxygen and carbon dioxide with the environment can take place through it (cutaneous respiration). Importantly, as almost all extant adult amphibians are carnivorous predators, sensitive inner ear structures and sticky tongues are common in this clade.

Evolution of tetrapods from fishes represented a significant change in body plan from one suited to organisms that respired and swam in water, to organisms that breathed air and moved onto land; these changes occurred over a span of 50 million years during the Devonian period.

Amniotes

The amniotes (reptiles, birds, and mammals) are distinguished from amphibians by their terrestrially adapted egg, which is protected by amniotic membranes. The evolution of amniotic membranes meant that the embryos of amniotes were provided with their own aquatic environment, which led to less dependence on water for development and thus allowed the amniotes to branch out into drier environments. This was a significant development that distinguished them from amphibians, which were restricted to moist environments due their shell-less eggs. Although the shells of various amniotic species vary significantly (from hard, to leathery, to using internal fertilization and development instead), they all allow retention of water.

The Amniotic Egg: a Closer Look

The amniotic egg is the key characteristic of amniotes. In amniotes that lay eggs, the shell of the egg provides protection for the developing embryo while being permeable enough to allow for the exchange of carbon dioxide and oxygen. The albumin, or egg white, provides the embryo with water and protein, whereas the fattier egg yolk is the energy supply for the embryo, as is the case with the eggs of many other animals, such as amphibians.

However, the eggs of amniotes contain three additional extra-embryonic membranes: the chorion, amnion, and allantois. Extra-embryonic membranes are membranes present in amniotic eggs that are not a part of the body of the developing embryo. While the inner amniotic membrane surrounds the embryo itself, the chorion surrounds the embryo and yolk sac. The chorion facilitates exchange of oxygen and carbon dioxide between the embryo and the egg's external environment. The amnion protects the embryo from mechanical shock and supports hydration. The allantois stores nitrogenous wastes produced by the embryo and also facilitates respiration. In mammals, membranes that are homologous to the extra-embryonic membranes in eggs are present in the placenta.

Reptiles

Reptiles are tetrapods, although some lineages have only vestigial structures since descending from four-limbed ancestors. Reptiles, even aquatic ones, lay eggs enclosed in shells on land. They usually reproduce sexually with internal fertilization, although some species show ovoviviparity (eggs remaining in the mother's body until they are ready to hatch) or viviparity (live-born offspring).

Crocodilians, such as this Siamese crocodile (Crocodylus siamensis), provide parental care for their offspring.

One of the key adaptations that permitted reptiles to live on land was the development of their scaly skin, containing the protein keratin and waxy lipids, which reduced water loss from the skin. This occlusive skin means that reptiles cannot use their skin for respiration, like amphibians, and thus all breathe with lungs.

Reptiles are ectotherms, animals whose main source of body heat comes from the environment. In addition to being ectothermic, reptiles are categorized as poikilotherms, or animals whose body temperatures vary rather than remain stable. Reptiles have behavioral adaptations to help regulate body temperature, such as basking in sunny places to warm up and finding shady spots or going underground to cool down. The advantage of ectothermy is low metabolic energy needed to survive (e.g., about 10% of a similarly-sized endotherm).

Birds

The most obvious characteristic that sets birds apart from other modern vertebrates is the presence of feathers, which are modified scales. Birds are endothermic, and because they fly, they require large amounts of energy, necessitating a high metabolic rate. Like mammals, which are also endothermic, birds have an insulating covering that keeps heat in the body: feathers. Specialized feathers called down feathers are especially insulating, trapping air in spaces between each feather to decrease the rate of heat loss. Certain parts of a bird's body are covered in down feathers, and the base of other feathers have a downy portion, whereas newly hatched birds are covered in down.

Feathers not only act as insulation but also allow for flight, enabling the lift and thrust necessary to become airborne. The feathers on a wing are flexible, so the collective feathers move and separate as air moves through them, reducing the drag on the wing.

Like the question of how flight evolved, the question of how endothermy evolved in birds still is unanswered. Feathers provide insulation, but this is only beneficial if body heat is being produced internally. Similarly, internal heat production is only viable if insulation is present to retain that heat. It has been suggested that one or the other– feathers or endothermy– evolved in response to some other selective pressure.

Birds as Modern Dinosaurs

The evolutionary history of birds is still somewhat unclear. Due to the fragility of bird bones, they do not fossilize as well as other vertebrates. Birds belong to a group called the archosaurs, which also includes crocodiles and dinosaurs.

Dinosaurs (including birds) are subdivided into two groups, the Saurischia ('lizard like') and the Ornithischia ('bird like'). Despite the names of these groups, most evidence suggests it was not the bird-like dinosaurs that gave rise to modern birds (but there is dissent and an alternative hypothesis has been proposed). Rather, the lizard-like dinosaurs gave rise to bipedal predators called theropods, which includes birds.

One important fossil of an animal intermediate to dinosaurs and birds is Archaeopteryx, which is from the Jurassic period. Archaeopteryx is important in establishing the relationship between birds and dinosaurs, because it is an intermediate fossil, meaning it has characteristics of both dinosaurs and birds. Some scientists propose classifying it as a bird, but others prefer to classify it as a dinosaur. The fossilized skeleton of Archaeopteryx looks like that of a dinosaur, and it had teeth whereas birds do not, but it also had feathers modified for flight, a trait associated only with birds among modern animals. Fossils of older feathered dinosaurs exist, but the feathers do not have the characteristics of flight feathers.

Mammals

Mammals are vertebrates that possess several defining characteristics, including certain features of the hair, the jaw and skeletal system, integument (skin), and internal anatomy.

The presence of hair is one of the most obvious signs of a mammal. Although it is not very extensive on certain species, such as whales, hair has many important functions for mammals. Mammals are endothermic, and hair provides insulation to retain heat generated by metabolic work. Hair traps a layer of air close to the body, retaining heat. Along with insulation, hair can serve as a sensory mechanism via specialized hairs called vibrissae, better known as whiskers. These attach to nerves that transmit information about sensation, which is particularly useful to nocturnal or burrowing mammals. Hair can also provide protective coloration or be part of social signaling.

The skeletal system of mammals possesses many unique features. Unlike other vertebrates, the lower jaw of mammals consists of just one bone, the dentary. The additional jaw bones found in other vertebrates have been modified to function in hearing and form bones in the middle ear (the malleus, incus, and stapes) and is one way of distinguishing fossil mammals from fossils of other synapsids.

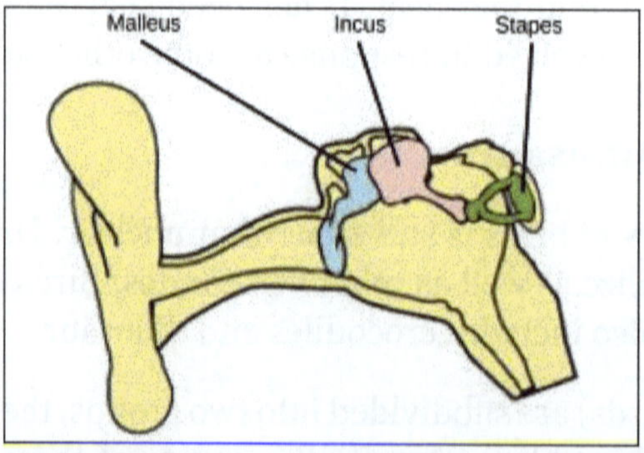

Bones of the mammalian inner ear are modified from bones of the jaw and skull.

The adductor muscle that closes the jaw is composed of two muscles in mammals the temporalis and the masseter. These allow side-to-side movement of the jaw, making chewing possible, which is unique to mammals. Most mammals have heterodont teeth, meaning that they have different types and shapes of teeth rather than just one type and shape of tooth.

In addition to defining mammary glands produce milk that is used to feed newborns, other specific glands appear to mark mammals uniquely:

- Sebaceous glands produce a lipid mixture called sebum that is secreted onto the hair and skin for water resistance and lubrication.
- Eccrine glands which produce sweat, may be limited to certain parts of the body or absent in some species, but when present, aid with thermoregulation.
- Apocrine glands, or scent glands, secrete substances that are used for chemical communication, such as in skunks.

Modern mammals belong to just three clades: eutherians (or placental mammals, like us), marsupials (eg., kangaroo, opossum), and monotremes (platypus and echidna).

2

Ichthyology: A Comprehensive Study

The discipline of zoology which focuses on the study of animals which belong to the class Pisces, commonly known as fish, is termed as ichthyology. The different categories of fish which are studied within this field include agnathan, chondrichthyes and osteichthyes. This chapter discusses in detail these types of fish to provide an extensive understanding of the subject.

Ichthyology is the science of animals that deals with fish. This field includes the study of fish growth, development, structure, characteristics, classification, geographical distribution, and the relationship of fish to their environment.

Pisces

The animals belonging to class-Pisces are commonly called fishes. They exclusively live in water.

The skin of fishes is covered with scales/plates, which helps these animals to live in water. The body may be long, laterally compressed and spindle shaped or dorsiventrally flattened and disc shaped. It usually consists of head, trunk and a muscular tail. The muscular tail and fins help them to swim in water and move from one places to another. Fishes are cold-blooded animals and their hearts have only two chambers – one auricle and one ventricle. Fishes obtain oxygen dissolved in water and breathe through gills. They are egg laying animals. Fertilization is external. There are many kinds of fishes. They have been broadly grouped under three categories.

Cyclostoma: The round mouthed fishes. Examples, The hagfish, the lamprey.

Chondrichthyes: The cartilaginous fishes. Examples Scoliodon (dog fish or the Indian Shark), stingray, electric ray (Torpedo).

Osteichthyes: The bony fishes. Examples, Labio rohita (Rohu), Hippocampus (Sea horse), Tuna, etc.

Body Structure

- Head, trunk (törzs) and tail - very streamlined.

- Skeletal muscles are organized in such a way that the animals can only swim in a forward direction.
- Locomotion with fins, which are membranes supported by rays or spines.

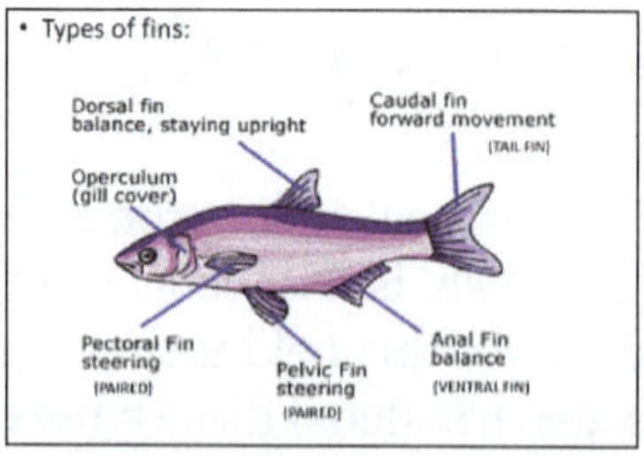

Swimming movements in a fish

Not all fish have all of the fins. Unpaired fins are the dorsal, caudal and anal fins. Paired fins are the pectoral and pelvic fins.

- Respiration occurs as water passes over the gills, which take up oxygen and release carbon dioxide.

The Gills of a Fish.

- Body covering: Scales (pikkelyek) and mucus (nyálka). Scales are thin, round discs that are modified bony plates. They provide mechanical protection. Scales overlap to reduce friction. They vary in number and size (some fish have no scales at all). Some scales can change colour, many are reflective. Mucus is produced by glands (mirigyek) in the skin and reduces friction.

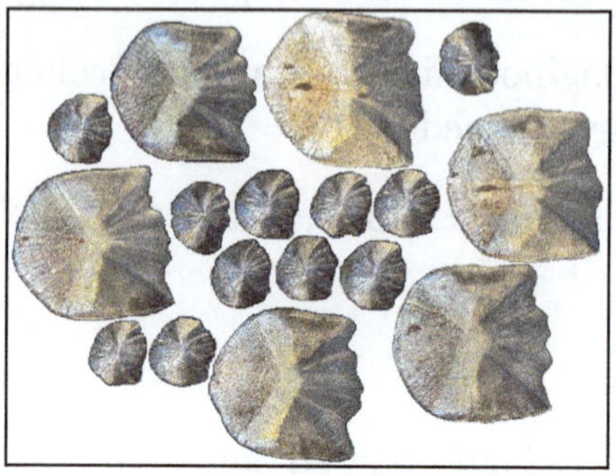

Various fish scales.

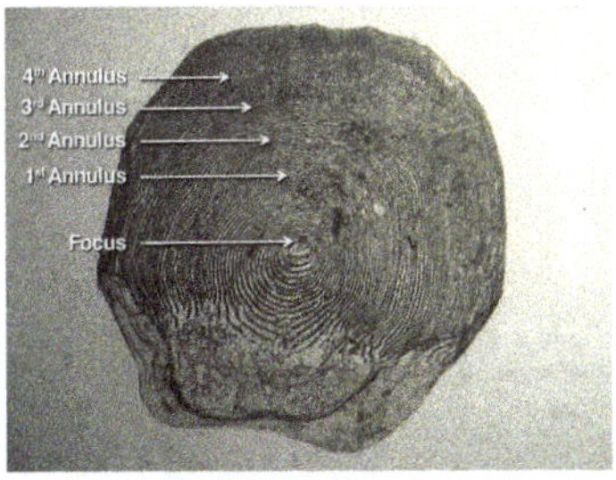

Scales can be used to age the fish, as rings form as the scales grow.

Shark scales tend to be ridged and pointy, given sharks a rough body surface.

- Sensory: Fish have a unique system called the *lateral line system* (oldalvonal rendszer). These are fluid-filled canals that run along the sides of the fish. They detect movement and vibrations in water.

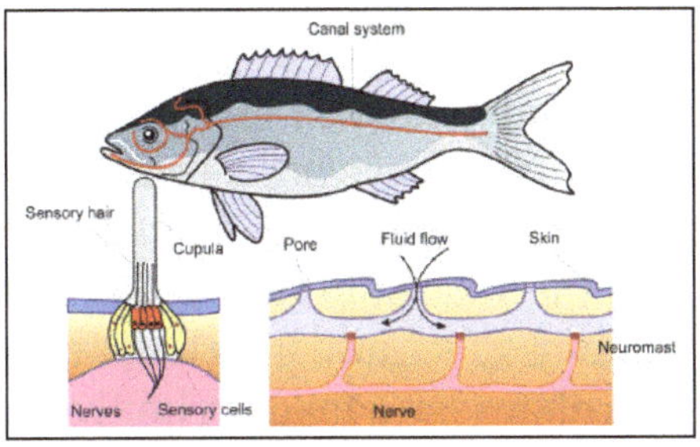

- Circulatory: 2 chambered heart, with one atrium (pitvar) and one ventricle (kamra). This is called single-circuit circulation because blood only passes through the heart once on its trip around the body.

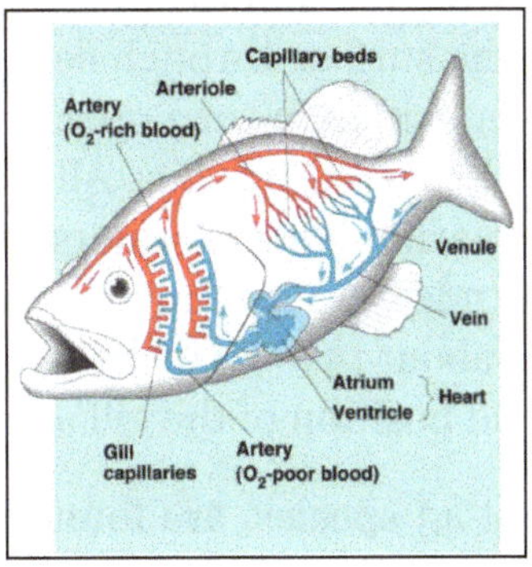

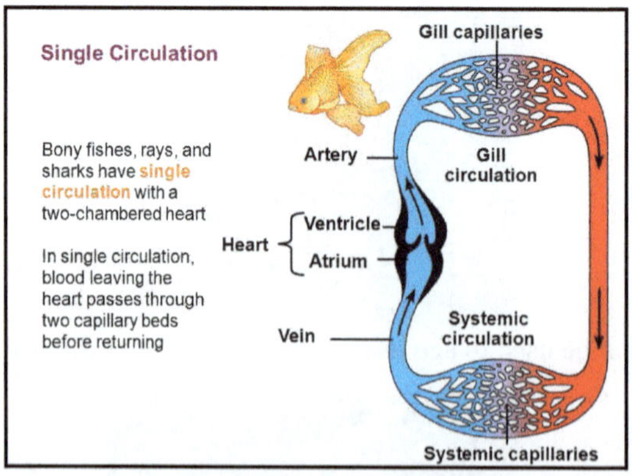

- Ectotherms/cold-blooded/poikilotherms: this means that their body temperature is not constant, it changes with the temperature of the surrounding environment.

Agnathan

Agnathan (superclass Agnatha) is any member of the group of primitive jawless fishes that includes the lampreys (order Petromyzoniformes), hagfishes (order Myxiniformes), and several extinct groups.

Hagfishes are minor pests of commercial food fisheries of the North Atlantic, but lampreys, because of their parasitic habit, have been a serious pest of food fisheries in the Great Lakes in North America, where they have reduced the numbers of lake trout and other species. Agnathans are otherwise of little economic importance. The group is of great evolutionary interest, however, because it includes the oldest known craniate fossils and because the living agnathans have many primitive characteristics.

General Features

The body of the hagfish is soft-skinned, scaleless, and nearly cylindrical, with a single nostril at the anterior end, overlying the mouth, and a low caudal fin around the tail. In some species the gills open to the surface through separate pores; however, in others the gills open to a common duct, which in turn opens to the surface through a single pore. The eyes are vestigial and covered by skin. All of the approximately 70 known species are restricted to cold, marine bottom waters at depths ranging from 10 metres (about 33 feet) in high latitudes to 1,300 metres (about 4,300 feet) in equatorial oceans. Adults are 40 to 80 cm (about 15 to 30 inches) long. All species are superficially similar except in the number and position of the gill apertures.

Lampreys, which number about 43 species, are found in cool, fresh, and coastal waters of all continents except Africa. All species are rather similar. The body is smooth, scaleless, and eel-shaped, with well-developed dorsal and caudal fins; the mouth is surrounded by a suctorial oral disk bearing horny teeth. The eyes are well developed, and the single nostril is on the top of the head. Lampreys possess seven pairs of external gills. Adults range from 15 to 100 cm (6 to 39 inches) long.

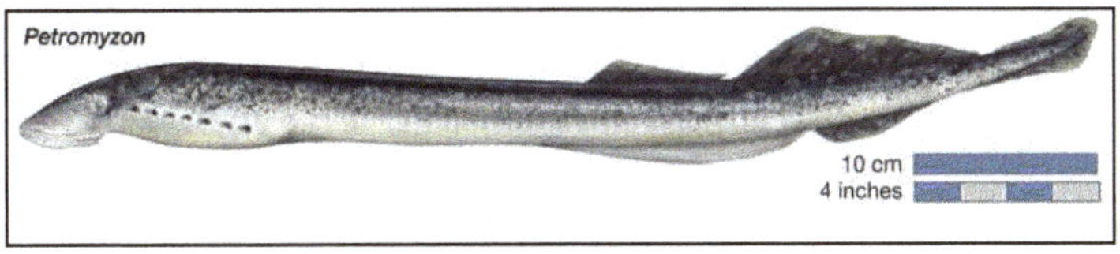

Lamprey (Petromyzon).

Although the gonad of a hagfish usually includes both ovary and testis, there is no evidence of either hermaphroditism (the reproductive organs of both sexes functioning in the same individual) or self-fertilization. A female produces a small number of tough-skinned yolk-filled eggs about 2 cm (0.8 inch) long that hatch into miniature adults.

Hagfishes locate their food by scent. Although some are known to eat fishes immobilized in nets, the best-studied species, Myxine glutinosa, normally feeds on soft-bodied invertebrates and larger dead animals. Myxine burrows into soft marine sediments and rests with only the tip of the head protruding. During respiration, water enters through the nostril and passes by a nasopharyngeal duct to the pharynx and gills. When stimulated by the scent of a dead fish, Myxine leaves its burrow and swims against the current. On making contact with the fish, Myxine coils around it and bites into it by protruding and retracting the comblike horny tooth plates on the floor of the mouth. Along each side of Myxine's body is a row of prominent glands that produce a gelatinous slime when it is disturbed.

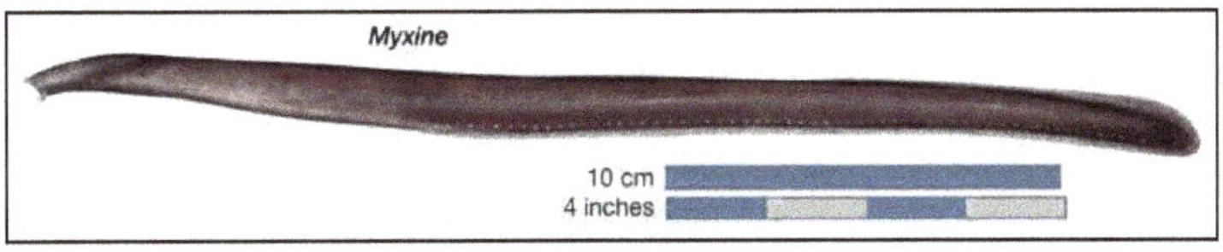

Hagfish (Myxine).

Lampreys mate in a nestlike depression excavated by the male in the gravel bed of a stream; the numerous eggs, about 1 mm (0.04 inch) in diameter, lodge in the gravel around the nest. The egg hatches into an ammocoete larva—a blind, wormlike animal that burrows in silt. The larva's mouth is overhung by a hoodlike upper lip that protrudes above the surface of the silt. A continuous stream of water passes in through the mouth and out through the seven pairs of gills. Microscopic plants, the food of the ammocoete, are filtered from this respiratory current by strands of mucus produced by the endostyle, which is a gland in the floor of the pharynx.

After about three years, when the ammocoete has grown to about 10 cm (about 4 inches), it undergoes a radical metamorphosis. The eyes complete their development; the upper lip becomes transformed into a suctorial oral disk; the endostyle changes into a thyroid gland; and the fins along the back increase in height. On completion of metamorphosis, a typical lamprey such as Petromyzon marinus migrates to the sea, where it feeds by attaching itself with its sucker to bony fishes. It rasps into the flesh with a toothed, tonguelike structure on the floor of the mouth. Saliva containing an anticoagulant facilitates the ingestion of blood and muscle tissue. On attaining full adult size, the lamprey ceases to feed, migrates upstream to a spawning ground, mates, and dies.

Members of several lamprey species do not migrate to sea but feed in fresh water. Petromyzon marinus dorsatus once seriously affected commercial fishing in the Great Lakes until measures were undertaken to control it. The brook lampreys do not feed after metamorphosis but mature sexually, reproduce, and die.

Form and Function

The elongated bodies of hagfishes and lampreys are adaptations to a burrowing habit, throughout life in the hagfish but only during the larval period in the lamprey. Considerable variation in body form occurs among extinct agnathans. Some Osteostraci, for example, possessed a flattened head shield and, although possessing a powerful swimming tail, appear to have been bottom dwellers. The size and shape of the mouth suggest that they were filter feeders. The laterally compressed, fishlike form of the anaspids (such as Pharyngolepis) indicates a free-swimming habit. The extinct heterostracans include obvious bottom dwellers (Drepanaspis) and others (Pteraspis) that were apparently adapted to mid-water, or nektonic, life. Some heterostracans, for example, had movable enamel plates inside the lower lip, probably to provide a biting or grazing mechanism. Osteostracans, anaspids, hagfishes, and lampreys have one median nostril, but heterostracans appear to have had two, one at each corner of the mouth.

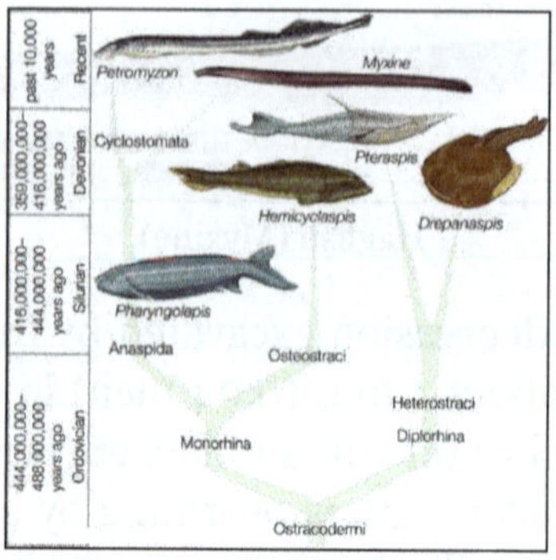

Systematic division of fossil and recent Agnatha.

Chondrichthyes

Chondrichthyan, (class Chondrichthyes), also called chondrichthian is any member of the diverse group of cartilaginous fishes that includes the sharks, skates, rays, and chimaeras. The class is one of the two great groups of living fishes, the other being the osteichthians, or bony fishes. The name Selachii is also sometimes used for the group containing the sharks.

Bull shark (Carcharhinus leucas).

Remora shark.

A remora (Echeneis naucrates) and its host, a zebra shark (Stegostoma fasciatum). By attaching itself to the shark, the remora is carried along by the shark, allowing the remora to travel to different areas without having to expend its own energy to swim. The shark is completely unaffected by the remora's presence.

Many unique structural, physiological, biochemical, and behavioral characters make these fishes of particular interest to scientists. These fishes are, in a sense, living fossils, for many of the living sharks and rays are assigned to the same genera as species that swam the Cretaceous seas over 100 million years ago. More than 400 species of sharks and about 500 species of rays are known. Although by any reckoning a successful group, the modern chondrichthyans number far fewer species than the more advanced bony fishes, or teleosts.

The danger some sharks and stingrays present to humans makes these animals fascinating and, at the same time, fearsome. Perhaps for this reason, they figure prominently in the folklore and art of many tropical peoples who depend on the sea. The danger from shark attack, while very real, is remarkably uncommon and easily sensationalized. Quite frequently, little attempt is made to distinguish between dangerous and harmless species.

General Features

Problems of Taxonomy

The name Selachii refers to a category of fishlike vertebrates characterized by a skeleton primarily composed of cartilage. Selachii are given a variety of treatments by ichthyologists. Some authorities consider Selachii to be a class or subclass that contains all the modern sharks and rays; other authorities restrict the use of the name to an order of modern sharks and certain extinct ancestral forms. Under the latter system, the rays (including the sawfishes, guitarfishes, electric rays, mantas, skates, and stingrays) are ranked separately.

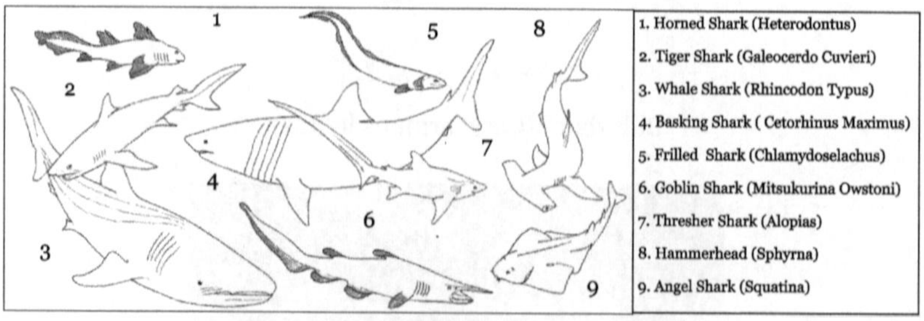

Body plans of representative sharks.

The chimaeras (Holocephali) bear many similarities to sharks and rays in skeletal structure, internal organs, and physiology. Ichthyologists commonly although not unanimously emphasize these similarities by grouping the modern and ancient sharks, rays, and chimaeras in the class Chondrichthyes, the cartilaginous fishes. Under this system, the sharks, skates, and rays are further grouped into one subclass, Elasmobranchii, and the chimaeras into another, Holocephali. Some authorities classify the elasmobranchs into one class (Selachii) and classify the chimaeras into another (Holocephali); however, assigning the two groups class rank implies a degree of distinctness equal to that of the amphibians (Amphibia), reptiles (Reptilia), birds (Aves), and mammals (Mammalia).

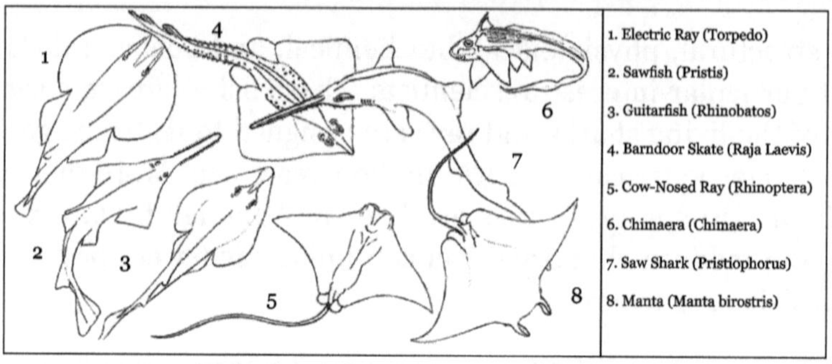

Body plans of representative Selachii.

Distribution and Abundance

The majority of sharks and rays are marine fishes, but many enter estuaries; some travel far up rivers, and a few are permanent residents of fresh water. Most species live in the relatively shallow waters of continental margins or around offshore islands; a few roam far out in the vast spaces of the oceans. Some live at great depths, in midwaters or on the bottom; others are surface swimmers or inhabit the bottom in shallow waters.

Once regarded as "trash" fish, sharks and rays are increasingly represented in the fisheries of most countries. With numbers of more highly valued bony fishes decreasingly rapidly, many fisheries are specifically targeting elasmobranchs as a primary commercial resource. Annual yields are as much as 750,000 metric tons (roughly 827,000 short tons), and most of this is sold fresh, dried, salted, or processed. This intense harvest is causing the reduction of many shark and ray populations and, in some cases, causing the commercial extinction of some species. One consequence of this depletion of these top-level predators may be a disruption of the food chain in marine ecosystems.

Food Habits

Sharks

All sharks are carnivorous and, with a few exceptions, have broad feeding preferences, governed largely by the size and availability of the prey. The recorded food of the tiger shark (Galeocerdo cuvier), for example, includes a wide variety of fishes (including other sharks, skates, and stingrays), sea turtles, birds, sea lions, crustaceans, squid, and even carrion such as dead dogs and garbage thrown from ships. Sleeper sharks (Somniosus), which occur mainly in polar and subpolar regions, are known to feed on fishes, small whales, squid, crabs, seals, and carrion from whaling stations. Many bottom-dwelling sharks, such as the smooth dogfishes (Triakis and Mustelus), take crabs, lobsters, and other crustaceans, as well as small fishes.

The three largest sharks, the whale shark (Rhincodon typus), the basking shark (Cetorhinus maximus), and the megamouth shark (Megachasma pelagios), resemble the baleen whales in feeding mode as well as in size. They feed exclusively or chiefly on minute passively drifting organisms (plankton). To remove these from the water and concentrate them, each of these species is equipped with a special straining apparatus analogous to baleen in whales. The basking shark and the megamouth shark have modified gill rakers, the whale shark elaborate spongy tissue supported by the gill arches. The whale shark also eats small, schooling fishes.

Saw sharks (Pristiophoridae) and sawfishes (Pristidae), though unrelated, both share a specialized mode of feeding that depends on the use of their long bladelike snout, or "saw." Equipped with sharp teeth on its sides, the saw is slashed from side to side, impaling, stunning, or cutting the prey fish. Saw sharks and sawfishes, like most other rays, are bottom inhabitants.

Thresher sharks (Alopias) feed on open-water schooling fishes, such as mackerel, herring, and bonito, and on squid. The long upper lobe of the tail, which may be half the total length of the shark, is used to herd the fish (sometimes by flailing the water surface) into a concentrated mass convenient for feeding. Thresher sharks have also been observed to stun larger fish with a rapid strike of the tail.

Most sharks and rays do not school. Individuals are normally solitary and usually come together only to exploit food resources or to mate. During these encounters, some species may show specific dominance structures, usually based on size. Some species, however, will travel in large schools segregated by size, a habit that protects smaller individuals from being eaten by larger ones. Still other species form sex-segregated schools where males and females live in slightly different habitats or depths. When potential prey is discovered, sharks circle it, appearing seemingly out of nowhere and frequently approaching from below. Feeding behaviour is stimulated by increasing numbers and rapid swimming, when three or more sharks appear in the presence of food. Activity soon progresses from tight circling to rapid crisscross passes. Biting habits vary with feeding methods and dentition. Sharks with teeth adapted for shearing and sawing are aided in biting by body motions that include rotation of the whole body, twisting movements of the head, and rapid vibrations of the head. As the shark comes into position, the jaws are protruded, erecting and locking the teeth into position. The bite is extremely powerful; a mako shark (Isurus), when attacking a swordfish too large to be swallowed whole, may remove the prey's tail with one bite. Under strong feeding stimuli, the sharks' excitement may intensify into what is termed a feeding frenzy, possibly the result of stimulatory overload, in which not only the prey but also injured members of the feeding pack are devoured.

In most cases, sharks locate food by smell, which is well developed in nearly all species. Sharks also possess other important senses that allow them to find food, and the importance of each sense varies between species. Their lateral line system, a series of sensory pores along the side of the body for detecting vibrations, allows sharks to detect vibrations in the water. Their network of ampullae allows them to sense weak electrical signals given off by prey, and their eyes are often acute enough to discriminate the size, shape, and colour of their prey. The sum of these senses working together makes a well-integrated system for finding prey.

Rays

The majority of batoid fishes (members of the order Batoidei such as rays and allies) are bottom dwellers, preying on other animals on or near the seafloor. Guitarfishes (Rhynchobatidae and Rhinobatidae), butterfly rays (Gymnuridae), eagle rays (Mylobatidae), and cow-nosed rays (Rhinopteridae) feed on invertebrates, principally mollusks and crustaceans. Whip-tailed rays (Dasyatidae) use their broad pectoral fins to dig shellfish from sand or mud. Skates (Rajidae) lie on the bottom, often partially buried, and rise in pursuit of such active prey as herring. Skates trap their victims by swimming over and then settling upon them, a practice facilitated by their habit of hunting at night.

Electric rays (Torpedinidae) are characteristically bottom fishes of sluggish habits. They feed on invertebrates and fish, which may be stunned by shocks produced from the formidable electric organs. With their electricity and widely extensible jaws, these rays are capable of taking very active fishes, such as flounder, eel, salmon, and dogfish. Shallow-water electric rays have been observed to trap fishes by suddenly raising the front of the body disk while keeping the margins down, thereby forming a cavity into which the prey is drawn by the powerful inrush of water.

Most of the myliobatoid rays (seven recognized families of the suborder Myliobatoidei [order Myliobatiformes], which includes all the typical rays) swim gracefully, with undulations of the broad winglike pectoral fins. Some species, especially the eagle rays, frequently swim near the surface and even jump clear of the water, skimming a short distance through the air.

Manta, or devil, rays (Mobulidae) swim mostly at or near the surface, progressing by flapping motions of the pectoral fins. Even the largest often leap clear of the water. In feeding, a manta moves through masses of macroplankton or schools of small fish, turning slowly from side to side and using the prominent cephalic fins, which project forward on each side of the mouth, to funnel the prey into the broad mouth.

Chimaeras and ghost sharks (Chimaeridae) dwell near the bottom in coastal and deep waters, to depths of at least 2,500 metres (about 8,000 feet). They are active at night, feeding almost exclusively on small invertebrates and fishes.

Reproductive Behaviour

Mature individuals of some species of sharks segregate by sex, coming together only during the mating season, when the males—at least those of the larger, more aggressive species—stop feeding. Segregation is a behavioral adaptation to protect the females. One principal courting activity used by the male to induce cooperation of the female in mating is the act of biting her and gripping her with his teeth. A male takes hold of a female in this way so that he can more easily insert a modified fin, called a clasper, into her cloaca. After mating, the sexes again separate. The pregnant females tend to keep apart from the other females of like size. As the time of parturition approaches, the pregnant females move to particular areas, which presumably have environmental properties especially suitable as nursery grounds. When giving birth to their young, they stop feeding, and, soon after parturition is completed, they depart.

Nursery areas vary with species. Some sharks—such as the bull shark (Carcharhinus leucas) and the sandbar shark (C. plumbeus)—use shallow waters of bays and estuaries; the silky shark (C. falciformis) uses the bottom far out on oceanic banks such as the Serrana Bank in the western Caribbean. The Atlantic spiny dogfish (Squalus acanthias) bears its young mostly during the winter, far out on the continental shelf of northeastern America, almost two years after mating.

A few skates that have been observed mating may be characteristic of other rays. The male seizes the female by biting the pectoral fin and presses his ventral surface against hers while inserting his clasper, or in some species both claspers, into her cloaca. Male skates have one to five rows of clawlike spines on the dorsal side of each pectoral fin. These are retractile in grooves of the skin and are used to hold the female during mating.

The eggs of skates in aquaria have been observed to be extruded in series, usually of two eggs at a time but sometimes one. Rest periods of one to five days occur between extrusions. A female of a European skate, Raja brachyura, laid 25 eggs over a 49-day period in the National Marine Aquarium, located in Plymouth, Eng.

Although the mating of chimaeroids has not been observed, it is generally presumed that the mode of copulation is similar to that of sharks and that the male's frontal spine and anterior appendage of the pelvic fins are probably used in securing the female. Two eggs are laid simultaneously, one from each oviduct. They are often carried for a relatively long period before being laid, several hours or even days, each egg protruding from the female for the greater part of its length.

Form and Function

Distinguishing Features

The elasmobranchs are fishlike vertebrates that differ from bony fishes in many respects. The skeleton is composed of cartilage and, although often calcified (especially in the vertebrae), lacks true bone (except in the roots of teeth). There are five to seven fully developed gill clefts, opening separately to the exterior. Most sharks and all rays have an opening behind each eye, called a spiracle, which is a modified first gill cleft. The dorsal fin or fins and fin spines are rigid, not erectile. Scales, if present, are structurally minute teeth, called dermal denticles, each consisting of a hollow cone of dentine surrounding a pulp cavity and covered externally by a layer of hard enamel-like substances called vitrodentine. The scales covering the skin do not grow throughout life, as they do in bony fishes, but have a limited size; new scales form between existing ones as the body grows. Certain other structures, such as the teeth edging the rostrum (beak) of sawfishes and saw sharks, the stinging spines of stingrays, and the teeth in the mouth, are structurally modified scales. The teeth, arranged in rows in the mouth, are not firmly attached to the jaws but are imbedded in a fibrous membrane lying over the jaws. When a tooth becomes broken, worn, or lost, it is replaced by one moving forward from the next row behind; at the base of the innermost row are rudimentary teeth and tooth buds that develop and move forward as needed. A spiral membranous fold (spiral valve) extends through the intestine of all sharks, rays, and chimaeras.

The rays differ externally from sharks in having the gill openings confined to the lower surface; the eyes of the rays are on the dorsal surface, and the edges of the pectoral fins are attached to the sides of the head in front of the gill openings. Some rays lack scales, and others are variously armed with thorns, tubercles, or prickles, all of which are modified scales; the tails of some have long, saw-toothed spines equipped with poison glands. In the sawfishes the snout is prolonged into a long, flat blade armed on either side with teeth. The electric rays have electric organs by which they can administer electric shocks to enemies or prey.

The chimaeras have only one external gill opening. In the adult the skin on each side of the head is smooth and lacks scales; the teeth consist of six pairs of grinding plates. The dorsal fin and spine are erectile. Like male sharks and rays, male chimaeras have claspers that serve to transfer sperm to the female, but in addition they have an erectile clasping device, the tantaculum, in front of each pelvic fin; most species have another such organ on top of the head.

Senses

Although sharks are often said to have a low order of intelligence, they, as well as rays and chimaeras, have survived successfully over a long period of geologic time. They are well equipped to locate prey and their own kind; to direct the course of their seasonal migrations; to discriminate specific localities; to respond to variations of temperature; to react to attractive or repelling substances in the water; and perhaps even to feel objects some distance away from them. They can see, hear, smell, taste, feel, and maintain their equilibrium. The roles of the sense organs have been studied in only a few species, principally sharks, and consequently remain imperfectly understood.

The sense of smell is highly developed and probably the principal means of locating prey and guiding the predator toward it. Given a favourable direction of current, sharks can detect incredibly minute concentrations—fractions of a part per million (that is, less than 1×10^{-6} parts)—of certain substances in the water, such as blood.

Although their eyes are structurally and functionally adapted for seeing, it is believed that their visual acuity in discerning the form and colour of an object varies between species. The importance of sight is relative to the habitat and feeding habits of each species. Fast predatory sharks tend to have more acute vision, and in some deep-diving species the eyes are well developed to maximize detection of ambient light.

The hearing apparatus, located in the auditory capsule of the cranium, includes a system of semicircular canals, which are responsible for maintaining equilibrium. Sharks seem to be remarkably sensitive to sounds of low frequency and to possess extraordinary faculty for directional hearing. Whether hearing is more sensitive than smell has not yet been established.

Sensory organs identified as taste buds are located on the floor, sides, and roof of the mouth and on the throat, as well as on the tongue. Experiments on several species of large sharks indicate that they do discriminate food types—preferring tuna, for example, to other fish species. Under some conditions, however, they become less fastidious, going into a feeding frenzy in which they attack anything, including others of their own kind.

Sensory organs located in the skin of all sharks, rays, and chimaeras receive a variety of information—vibrations of low frequencies, temperature, salinity, pressure, and minute electrical stimuli, such as that produced by another fish in the vicinity. These sensory organs are located in the lateral line system, in groups of pores called ampullar organs, found on the head, snout, and around the jaws, which detect electrical impulses.

Salt and Water Balance

Most marine vertebrates maintain lower concentrations of salts and other chemicals in their blood than are found in seawater. As a result, these animals face a continuous problem of water loss to the environment, because of the tendency of water to move through membranes from regions of low salt concentration to regions of higher concentration. The marine cartilaginous fishes differ from almost all of the bony fishes (except the coelacanths and aestivating lungfishes) in being able to reabsorb in the renal (kidney) tubules most of their nitrogenous waste products (urea and trimethylamine oxide) and to accumulate these products in their tissues and blood, an ability termed the urea retention habitus. The concentration within the body thus exceeds that of the surrounding seawater, and water moves into the body with no expenditure of energy. When any of these fishes moves into fresh water, as many do, the urine flow to the outside increases; hence, the concentration of urea in the blood decreases. In the sawfish, for example, the increase of urine output is more than twentyfold; the blood urea concentration decreases to less than one-third the amount observed in marine forms. Purely freshwater elasmobranchs, such as the stingrays of the Orinoco and Amazon drainage systems, seem to lack the urea retention habitus.

Respiration

Sharks breathe chiefly by opening the mouth while expanding the mouth-throat (bucco-pharyngeal) cavity and contracting the gill pouches to close the gill slits. With the mouth closed, they contract the bucco-pharyngeal cavity while dilating the gill pouches, thus drawing the water over the gills where the exchange of oxygen and carbon dioxide takes place. Then, with the mouth still closed, they contract the bucco-pharyngeal cavity and gill pouches, and the gill slits are opened to expel the water.

Most of the rays, on the other hand, take in water chiefly through the spiracles; these then close by contraction at their anterior margins, which bear rudimentary gill filaments and a spiracular valve. Folds of membrane on the roof and floor of the mouth prevent the water from passing down the throat and direct it to the gill openings. Skates, which sometimes hold the lower surface of the head slightly above the bottom, may inhale some water through the mouth; mantas, which have small spiracles and live near the surface, respire chiefly through the mouth. Skates, stingrays, guitarfishes, and angel sharks frequently reverse the direction of flow through the spiracles, apparently to clear them of foreign matter.

Chimaeras take in water chiefly through the nostrils, keeping the mouth closed for the most part. The water reaches the mouth primarily through grooves leading there from the nostrils.

Reproduction and Development

All species of sharks, rays, and chimaeras produce large yolk-rich eggs. These are fertilized internally, for which the males are equipped with two copulatory organs called claspers along the inner edges of the pelvic fins. Each clasper has a groove for guidance of sperm. The few published descriptions of mating sharks and rays are probably characteristic of the entire group. The male grasps one of the female's pectoral fins with his teeth to hold her in position as he inserts a clasper through a cavity (cloaca) and into a tube (oviduct). Males of most species probably use only one clasper at a time. The sperm travel to the anterior end of the oviduct, where they fertilize the eggs. The eggs then move down the oviduct past the shell gland, where they are covered by a shell or capsule.

Some of the sharks, probably all the skates, possibly some of the guitarfishes, and all of the chimaeras are oviparous (egg-laying species). The eggs are enveloped in a horny shell, usually equipped with tendrils for coiling around solid objects or with spikelike projections for anchoring in mud or sand. The egg cases of most species are more or less pillow-shaped; those of the horned sharks (Heterodontus francisci) are screw-shaped with a spiral flange. The eggs of chimaeras are elliptic, spindle-shaped, or tadpole-shaped and open to the exterior through pores and slits that permit entrance of water during incubation. An egg of the whale shark found in the Gulf of Mexico measured 30 cm (12 inches) long by about 14 cm (5.5 inches) wide and was 8 cm (3 inches) thick. Protected by the shell and nourished by the abundant yolk, the embryo of an oviparous species develops for 18 to 59 weeks before hatching.

The majority of sharks and rays other than the skates are ovoviviparous (that is, the egg hatches within the mother). In this case, the egg is first coated in the shell gland with a temporary membranous capsule that lasts only during early development. After emerging from its capsule, the embryo remains in the oviduct of the mother, nourished by the yolk sac to which it remains attached. Embryos of some ovoviviparous sharks, notably the porbeagle (Lamna nasus), the mako (Isurus oxyrinchus), and the sand shark (Odontaspis taurus), ingest yolks of other eggs and even other embryos within the oviduct of the mother after the contents of their own yolk sacs are exhausted. In the majority of ovoviviparous sharks and rays, organically rich uterine secretions provide supplemental nourishment, which is absorbed by the yolk sac and in many cases by appendages borne on its stalk. In some genera of rays, vascular filaments producing these secretions extend through the spiracles and into the digestive tract of the embryos.

Several shark species are viviparous—that is, the yolk sac develops folds and projections that interdigitate with corresponding folds of the uterine wall, thus forming a yolk-sac placenta through which nutrient material is passed from the mother.

Growth

Growth of a few shark species has been measured or estimated by the differences in length at the times of tagging and recapturing specimens. Growth is also measured by the statistical analysis of the length in systematically collected samples, by the space between concentric circles on the centra of the vertebrae, and by periodic measurements of specimens kept in aquariums. All studies indicate a slow growth rate. During the 10 years between birth and maturity, male Atlantic spiny dogfish grow an average of 47 cm (19 inches) and females 67 cm (26 inches). The Greenland shark (Somniosus microcephalus), which attains 6.5 metres (21 feet) or more (although rarely taken larger than about 4 metres [13 feet]), grows only about 7.5 mm (about 0.3 inch) per year. The annual growth increments of tagged juvenile whitetip reef and Galapagos sharks, both species that become at least 2.5 metres (8 feet) long, were found to be 31 to 54 mm (1 to 2 inches) and 41 mm (about 1.5 inches), respectively. The Australian school shark (Galeorhinus australis) grows about 80 mm (3 inches) in its first year and about 30 mm (1 inch) in its 12th year. By its 22nd year, it is estimated to be approaching its maximum length of 1.6 metres (about 5 feet).

The disk of the eastern Pacific round stingray (Urolophus halleri) increases in width on the average from 75 mm (3 inches) at birth to 150 mm (6 inches) when mature (that is, at 2.6 years old). In the next five years it grows about 60 mm (about 2.4 inches) more toward its maximum recorded width of 25 cm (10 inches) in males or 31 cm (12.25 inches) in females. The males of European thornback rays (Raja clavata) are about 50 cm (20 inches) wide when they reach first maturity, about seven years after birth; females are 60 to 70 cm (24 to 28 inches) at first maturity, nine years after birth.

Osteichthyes

Osteichthyes, known as the bony fish, are a taxonomic class (or superclass) of fish and the largest class of vertebrates in existence today. With over 26,000 species, they comprise over 95 percent of all fish species. The Osteichthyes include the ray-finned fish (subclass or class Actinopterygii) and lobe finned fish (subclass or class Sarcopterygii). However, some taxonomic schemes do not consider Osteichthyes a formal taxonomic group, as it is not monophyletic (Nelson 2006), although the term is retained for vernacular use for those actinopterygians and sarcopterygians conventionally termed fishes.

The members of Osteichthyes represent an extraordinary diversity of forms, including the "living fossil" coelacanths, long "sea-serpent" oarfish, tiny cichlids, whiskered catfish, colorful clownfish, voracious piranhas, unique seahorses, and so forth. This diversity of forms, behaviors, and colors provides aesthetic joy to humans, as well as a variety of practical uses—food, sportfishing, aquarium fish, and so forth.

Although lampreys, sharks, and rays are also fish, some individuals commonly use the term only in reference to the jawed bony fish, that is, members of Osteichthyes.

Classification

Common Clownfish (Amphiprion ocellaris) in their magnificent sea anemone (Heteractis magnifica) home.

Fish are a very diverse assemblage, so much so that the term *fish* is itself more one of convenience than a taxonomic rank. In general, a fish can be defined as a poikilothermic (cold-blooded), water dwelling vertebrate with gills throughout life, and limbs—if present—in the form of fins. Poikilothermic refers to the fact that the internal temperatures of fish vary, often matching the ambient temperature of the environment. (However, some fish, such as the blue-fin tuna, can flourish at significantly higher internal temperatures than the surrounding water.)

Modern fish are conventionally divided into the *jawless fish* (class or superclass Agnatha, with about 75 species including lampreys and hagfish), the *cartilaginous fish* (class Chondrichthyes, with about 800 species including sharks and rays), and the *bony fish* (class or superclass *Osteichthyes*, with over 26,000 species). Some individuals do not include as fish those members of Agnatha or Chondrichthyes.

With over 27,000 known extant (living) species, fish constitute more than one-half of the total vertebrate species. The vast majority of fish are members of Osteichthyes. Nelson recognizes 45 orders, 435 families, 4079 genera, and 23,689 species of osteichthyans (bony fishes).

Within the Osteichthyes, two extant subclasses (or classes) are generally recognized: the Sarcopterygii (lobe-finned fish) and the Actinopterygii (ray-finned or spiny-finned fish). Most bony-fish belong to the Actinopterygii; there are only eight living species of lobe finned fish, including the lungfish (6 extant species) and coelacanths (2 extant species). Some species of lobe-finned fish have jointed bones. Actinopterygii, or ray-finned fish, include the most familiar fish, such as sturgeons, gars, eels, carp, herrings, anchovies, catfishes, goldfishes, piranhas, oarfish, seahorses, bass, cichlids, pickerel, salmon, and trout.

The Actinopterygii are generally classified into two groups—the Chondrostei and the Neopterygii—the latter of which includes the Teleostei (modern bony fishes), a classification into which most fish fit today. The Sarcopterygii are traditionally divided into the lungfish and coelacanths.

While Osteichthyes traditionally is treated as a class of vertebrates, with subclasses Actinopterygii and Sarcopterygii, some newer schemes divide this group into separate classes. Nelson, for example, lists Actinopterygii and Sarcopterygii as classes.

However, while using Osteichthyes as vernacular for actinopterygians and those sarcopterygians conventionally termed fishes, Nelson prefers not to use the term as a formal taxon as it "is clearly not a monophyletic group." Rather, Nelson uses the term "Euteleostomi" for Sacropterygii and Actinopterygii, while Nelson places Actinopterygii and Sarcopterygii (along with the extinct Acanthodii) in grade Teleostomi. Furthermore, Nelson places the tetrapods (mammals, birds, reptiles, amphibians) in class Sarcopterygii in order to make this a monophyletic group.

Characteristics

Osteichthyes are characterized by a relatively stable pattern of cranial bones, rooted teeth, and medial insertion of mandibular muscle in lower jaw. The head and pectoral girdles are covered with large dermal bones. The eyeball is supported by a sclerotic ring of four small bones, but this characteristic is missing or modified in many modern species.

The labyrinth in the inner ear contains large otoliths. The braincase, or neurocranium, is frequently divided into anterior and posterior sections divided by fissure.

Osteichthyes have a lung or swim bladder. They do not have fin spines, but instead support the fin with lepidotrichia (bone fin rays). They also have an operculum, which helps them breathe without having to swim.

The fin-limbs of sarcopterygiians show such a strong similarity to the expected ancestral form of tetrapod limbs that they have been universally considered the direct ancestors of tetrapods (four-legged vertebrates) in the scientific literature.

Replacement Bone

One of the best-known innovations of the osteichthians is endochondral bone or "replacement" bone, i.e. bone ossified internally, by replacement of cartilage, as well as perichondrally, as "spongy bone."

In the more general vertebrates, there are various types of calcified tissue: dentine, enamel (or "enameloids"), and bone, plus variants characterized by their ontogeny, chemistry, form and location. But endochondral bone is unique because it begins life as cartilage.

In more basal vertebrates, cartilaginous structures can become superficially calcified. However, in osteichthians, the circulatory system invades the cartilaginous matrix. This permits the local osteoblasts (bone-forming cells) to continue bone formation within the cartilage and also recruits additional, circulating osteoblasts. Other cells gradually eat away at the surrounding cartilage. The net result is that the cartilage is replaced from within by a somewhat irregular vascularized network of bone. Structurally, the effect is to create a relatively lightweight, flexible, "spongy" bone interior, surrounded by an outline of dense, lamellar periostial bone. Since this bone now surrounds other bone, rather than cartilage, it is referred to as periostial rather than perichondral. This is the unique endochondral bone from which the osteichthians derived their name, as well as many structural advantages.

However useful endochondral bone may be, it is also much heavier and less flexible than cartilage. Thus, many modern osteichian groups, including the extremely successful teleosts, have evolved away from extensive use of endochondral bone.

Examples: The ocean sunfish is the most massive bony fish in the world (but not the longest one; that honor goes to the oarfish). Specimens of ocean sunfish have been observed up to 3.33 m (11 ft) in length and weighing up to 2,300 kg (5,070 lb). Other very large bony fish include the Atlantic blue marlin, some specimens of which have been recorded as in excess of 820 kilograms (1,807.4 lb.), the black marlin, and some sturgeon species.

References

- Pisces, Biology: knowledgeuniverseonline.com, Retrieved 4 February, 2019
- Agnathan, animal: britannica.com, Retrieved 11 April, 2019
- Chondrichthian, animal: britannica.com, Retrieved 18 January, 2019
- Osteichthyes: newworldencyclopedia.org, Retrieved 8 June, 2019

3

Understanding Mammalogy

The vertebrate animals which belong to the class Mammalia are called mammals. The study of mammals is called mammalogy. Some of the different classifications within mammals are Theria and Prototheria. The topics elaborated in this chapter will help in gaining a better perspective about these types of mammals.

Mammalogy is the study of mammals. This is one of the newer classes of animals and is believed to be only 200 millions year old. There are approximately 4500 species of mammals. Despite the fact they represent a very small fraction of the world's animals, mammals are among the world's best known species. New species of mammal are still being discovered although many more go extinct than are discovered.

Mammal

Mammal, (class mammalia) is any member of the group of vertebrate animals in which the young are nourished with milk from special mammary glands of the mother. In addition to these characteristic milk glands, mammals are distinguished by several other unique features. Hair is a typical mammalian feature, although in many whales it has disappeared except in the fetal stage. The mammalian lower jaw is hinged directly to the skull, instead of through a separate bone (the quadrate) as in all other vertebrates. A chain of three tiny bones transmits sound waves across the middle ear. A muscular diaphragm separates the heart and the lungs from the abdominal cavity. Only the left aortic arch persists. (In birds the right aortic arch persists; in reptiles, amphibians, and fishes both arches are retained.) Mature red blood cells (erythrocytes) in all mammals lack a nucleus; all other vertebrates have nucleated red blood cells.

Except for the monotremes (an egg-laying order of mammals comprising echidnas and the duck-billed platypus), all mammals are viviparous—they bear live young. In the placental mammals (which have a placenta to facilitate nutrient and waste exchange between the mother and the developing fetus), the young are carried within the mother's womb, reaching a relatively advanced stage of development before birth. In the marsupials (e.g., kangaroos, opossums, and wallabies), the newborns are incompletely developed at birth and continue to develop outside the womb, attaching themselves to the female's body in the area of her mammary glands. Some marsupials have a pouchlike structure or fold, the marsupium, that shelters the suckling young.

The class Mammalia is worldwide in distribution. It has been said that mammals have a wider distribution and are more adaptable than any other single class of animals, with the exception of certain less-complex forms such as arachnids and insects. This versatility in exploiting Earth is attributed in large part to the ability of mammals to regulate their body temperatures and internal environment both in excessive heat and aridity and in severe cold.

General Features

Diversity

The evolution of the class Mammalia has produced tremendous diversity in form and habit. Living kinds range in size from a bat weighing less than a gram and tiny shrews weighing but a few grams to the largest animal that has ever lived, the blue whale, which reaches a length of more than 30 metres (100 feet) and a weight of 180 metric tons (nearly 200 short [U.S.] tons). Every major habitat has been exploited by mammals that swim, fly, run, burrow, glide, or climb.

Okapi (Okapia johnstoni).

There are more than 5,500 species of living mammals, arranged in about 125 families and as many as 27–29 orders (familial and ordinal groupings sometimes vary among authorities). The rodents (order Rodentia) are the most numerous of existing mammals, in both number of species and number of individuals, and are one of the most diverse of living lineages. In contrast, the order Tubulidentata is represented by a single living species, the aardvark. The Uranotheria (elephants and their kin) and Perissodactyla (horses, rhinoceroses, and their kin) are examples of orders in which far greater diversity occurred in the late Paleogene and Neogene periods (about 30 million to about 3 million years ago) than today.

The greatest present-day diversity is seen in continental tropical regions, although members of the class Mammalia live on (or in seas adjacent to) all major landmasses. Mammals can also be found on many oceanic islands, which are principally, but by no means exclusively, inhabited by bats. Major regional faunas can be identified; these resulted in large part from evolution in comparative isolation of stocks of early mammals that reached these areas. South America (the Neotropics), for example, was separated from North America (the Nearctic) from about 65 million to 2.5 million years ago. Mammalian groups that had reached South America before the break between the continents, or some that "island-hopped" after the break, evolved independently from relatives that remained in North America. Some of the latter became extinct as the result of competition with more advanced groups, whereas those in South America flourished, some radiating to the extent that they have successfully competed with invaders since the rejoining of the two continents. Australia provides a parallel case of early isolation and adaptive radiation of mammals (specifically the monotremes and marsupials), although it differs in that Australia was not later connected to any other landmass. The placental mammals that reached Australia (rodents and bats) evidently did so by island-hopping long after the adaptive radiation of the mammals isolated early on.

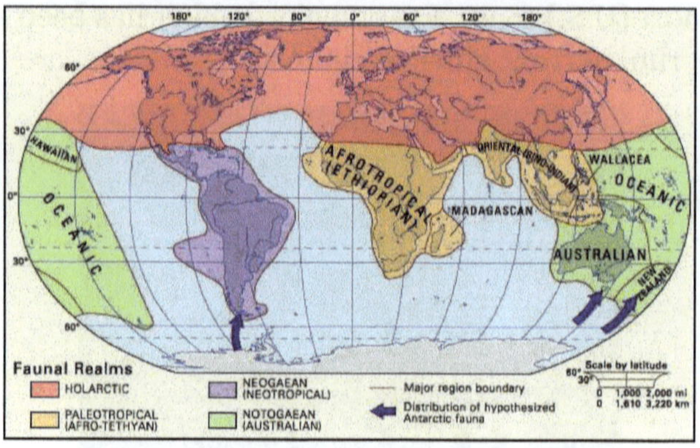

Faunal realms and major regions of the world.

In contrast, North America and Eurasia (the Palearctic) are separate landmasses but have closely related faunas as the result of having been connected several times during the Pleistocene Epoch (2.6 million to 11,700 years ago) and earlier across the Bering Strait. Their faunas frequently are thought of as representing not two distinct units but one, related to such a degree that a single name, Holarctic, is applied to it.

Reproduction

Estrus and Other Cycles

In reproductively mature female mammals, an interaction of hormones from the pituitary gland and the ovaries produces a phenomenon known as the estrous cycle. Estrus, or "heat," typically coincides with ovulation, and during this time the female is receptive to the male. Estrus is preceded by proestrus, during which ovarian follicles mature under the influence of a follicle-stimulating hormone from the anterior pituitary. The follicular cells produce estrogen, a hormone that stimulates proliferation of the uterine lining, or endometrium. Following ovulation, in late estrus, the ruptured ovarian follicle forms a temporary endocrine gland known as the corpus luteum. Another hormone, progesterone, secreted by the corpus luteum, causes the endometrium to become quiescent and ready for implantation of the developing egg (blastocyst), should fertilization occur. In members of the infraclass Eutheria (placental mammals), the placenta, as well as transmitting nourishment to the embryo, has an endocrine function, producing hormones that maintain the endometrium throughout gestation.

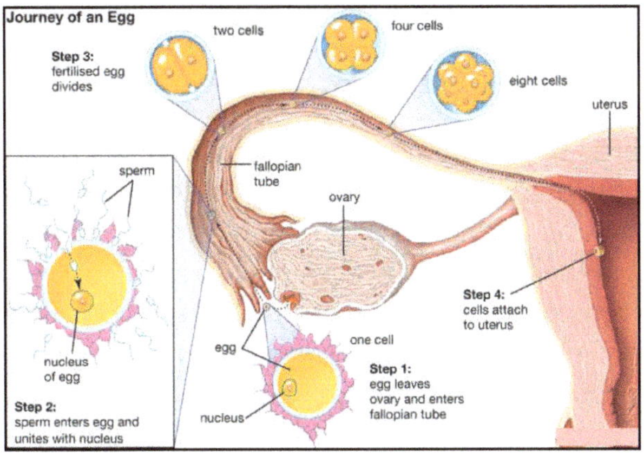

Journey of a fertilized egg: The journey of a fertilized egg in a woman.
In mammals, eggs are released by the ovaries.

If fertilization and implantation do not occur, a phase termed metestrus ensues, in which the reproductive tract assumes its normal condition. Metestrus may be followed by anestrus, a nonreproductive period characterized by quiescence or involution of the reproductive tract. On the other hand, anestrus may be followed by a brief quiescent period (diestrus) and another preparatory proestrus phase. Mammals that breed only once a year are termed monestrous and exhibit a long anestrus; those that breed more than once a year are termed polyestrous. In many polyestrous species the estrous cycle ceases during gestation and lactation (milk production), but some rodents have a postpartum estrus and mate immediately after giving birth.

The menstrual cycle of higher primates is derived from the estrous cycle but differs from estrus in that when progesterone secretion from the corpus luteum ceases, in the absence of fertilization, the uterine lining is sloughed. In anthropoids other than humans, a distinct period of "heat" occurs around the time of ovulation.

Monotremes lay shelled eggs, but the ovarian cycle is similar to that of other mammals. The eggs are predominantly yolk (telolecithal), like those of reptiles and birds. Young monotremes hatch in a relatively early stage of development and are dependent upon the parent (altricial). They reach sexual maturity in about one year.

The reproduction of marsupials differs from that of placentals in that the uterine wall is not specialized for the implantation of embryos. The period of intrauterine development varies from about 8 to 40 days. After this period the young migrate through the vagina to attach to the teats for further development. The pouch, or marsupium, is variously structured. Many species, such as kangaroos and opossums, have a single well-developed pouch; in some phalangerids (cuscuses and brush-tailed possums), the pouch is compartmented, with a single teat in each compartment. The South American caenolestids, or rat opossums, have no marsupium. The young of most marsupials depend on maternal care through the pouch for considerable periods, 13 to 14 weeks in the North American, or Virginia opossum (Didelphis virginiana). Young koalas are carried in the pouch for nearly 8 months, kangaroos to 10 months.

Implantation, Gestation and Birth

Reproductive patterns in placental mammals are diverse, but in all cases a secretory phase is present in the uterine cycle, and the endometrium is maintained by secretions of progesterone from the corpus luteum. The blastocyst implants in the uterine wall. Villi are embedded in the lining of the uterus. The resulting complex of embryonic and maternal tissues is a true placenta. The uterine lining may be shed with the fetal membranes as "afterbirth" (a condition called deciduate) or may be resorbed by the female (nondeciduate). Placentas have been classified on the basis of the relationship between maternal and embryonic tissues. In the simplest nondeciduate placental arrangement, the chorionic villi are in contact with uterine epithelium (the inner surface layer). In the "intimate deciduous" types, seen in primates, bats, insectivores, and rodents, the capillary endothelium (the layer containing minute blood vessels) of the uterine wall breaks down, and chorionic epithelium is in direct contact with maternal blood. In advanced stages of pregnancy in rabbits, even the chorionic epithelium is eroded, and the embryonic endothelium contacts the maternal blood supply. In no case, however, is there actual exchange of blood between mother and fetus; nutrients and gases must still pass through the walls of the fetal blood vessels.

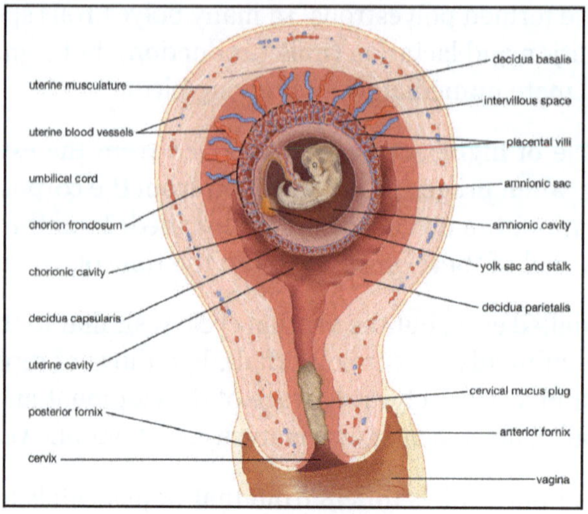

Embryonic Development.

The period of intrauterine development, or gestation, varies widely among eutherians, generally depending on the size of the animal but also influenced by the number of young per litter and the condition of young at birth. The gestation period of the golden hamster is about 2 weeks, whereas that of the blue whale is 11 months and that of the African elephant 21 to 22 months.

At birth the young may be well-developed and able to move about at once (precocial), or they may be blind, hairless, and essentially helpless (altricial). In general, precocial young are born after a relatively long gestation period and in a small litter. Hares and many large grazing mammals bear precocial offspring. Rabbits, carnivores, and most rodents bear altricial young.

After birth young mammals are nourished by milk secreted by the mammary glands of the female. The development of milk-producing tissue in the female mammae is triggered by conception, and the stimulation of suckling the newborn prompts copious lactation. In therians (marsupials and placentals) the glands open through specialized nipples. The newborn young of marsupials are unable to suckle, and milk is "pumped" to the young by the mother.

Milk consists of fat, protein (especially casein), and lactose (milk sugar), as well as vitamins and salts. The actual composition of milk of mammals varies widely among species. The milk of whales and seals is some 12 times as rich in fats and 4 times as rich in protein as that of domestic cows but contains almost no sugar. Milk provides an efficient energy source for the rapid growth of young mammals; the weight at birth of some marine mammals doubles in five days.

Behaviour

Social Behaviour

The dependence of the young mammal on its mother for nourishment has made possible a period of training. Such training permits the nongenetic transfer of information between generations. The ability of young mammals to learn from the experience of their elders has allowed a behavioral plasticity unknown in any other group of organisms and has been a primary reason for the evolutionary success of mammals. The possibility of training is one of the factors that has made increased brain complexity a selective advantage. Increased associational potential and memory extend the possibility of learning from experience, and the individual can make adaptive behavioral responses to environmental change. Individual response to short-term change is far more efficient than genetic response.

Some types of mammals are solitary except for brief periods when the female is in estrus. Others, however, form social groups. Such groups may be reproductive or defensive, or they may serve both functions. Within the social group, the hierarchy may be maintained through physical combat between individuals, but in many cases stereotyped patterns of behaviour evolve to displace actual combat, thereby conserving energy while maintaining the social structure.

Leopard (Panthera pardus) grooming her cub in the Masai Mara National Reserve, Kenya.

A pronounced difference between sexes (sexual dimorphism) is frequently extreme in social mammals. In large part this is because dominant males tend to be those that are largest or best-armed. Dominant males also tend to have priority in mating or may even have exclusive responsibility for mating within a "harem." Rapid evolution of secondary sexual characteristics, including size, can take place in a species with such a social structure.

A complex behaviour termed "play" frequently occurs between siblings, between members of an age class, or between parent and offspring. Play extends the period of maternal training and is especially important in social species, providing an opportunity to learn behaviour appropriate to the maintenance of dominance.

Territoriality

That area covered by an individual in its general activity is frequently termed the home range. A territory is a part of the home range defended against other members of the same species. As a generalization it may be said that territoriality is more important in the behaviour of birds than of mammals, but data for the latter are available primarily for diurnal species. Frequently territories of mammals are "marked," either with urine or with secretions of specialized glands, as in lemurs. This form of territorial labeling is less evident to humans than the singing or visual displays of birds. Many mammals that do not maintain territories per se nevertheless will not permit unlimited crowding and will fight to maintain individual distance. Such mechanisms result in more economical spacing of individuals over the available habitat.

Yellowstone wolf pack: Wolves in a pack in Yellowstone National Park, Wolves are territorial mammals, and members of the pack frequently mark the limits of the pack's territory with urine and secretions from specialized glands. Howling also serves to alert other animals to the pack's land claims.

Ecology

Response to Environmental Cycles

Mammals may react to environmental extremes with acclimatization, compensatory behaviour, or physiological specialization. One way for a mammal to endure stressful environmental conditions is to become dormant. Dormancy is the general term that relates to the reduced metabolic activity adopted by many organisms under conditions of environmental stress. Dormancy is differentiated from sleep, which is not necessarily a response to environmental stess but rather occurs as part of an organism's daily rest cycle. Physiological responses to adverse conditions include torpor, hibernation (in winter), and estivation (in summer). Torpor is a type of dormancy that may occur in the daily cycle or during unfavourable weather; short-term torpor is generally economical only for small mammals that can cool and warm rapidly. The body temperature of most temperate-zone bats drops near that of the ambient air whenever the animal sleeps. The winter dormancy of bears at high latitudes is an analogous phenomenon and cannot be considered true hibernation.

Hibernating bats: A cluster of Virginia big-eared bats (Corynorhinus townsendii virginianus) hibernating in Hellhole Cave, Pendleton county, West Virginia, U.S.

Strictly speaking, hibernation only occurs in warm-blooded vertebrates. True hibernation involves physiological regulation to minimize the expenditure of energy. The body temperature is lowered, and breathing may be slowed to as low as 1 percent of the rate in an active individual. There is a corresponding slowing of circulation and typically a reduction in the peripheral blood supply. When the body temperature nears the freezing point, spontaneous arousal occurs, although other kinds of stimuli generally elicit only a very slow response. In mammals that exhibit winter dormancy (such as bears, skunks, and raccoons), arousal may be quite rapid. Hibernation has evidently originated independently in a number of mammalian lines, and the comparative physiology of this complex phenomenon is only now beginning to be understood.

Inactivity in response to adverse summer conditions (heat, drought, lack of food) is termed estivation. Estivation in some species is simply prolonged rest, usually in a favourable microhabitat; in other species estivating mammals regulate their metabolism, although the effects are typically not as pronounced as in hibernation.

Behavioral response to adverse conditions may involve the selection or construction of a suitable microhabitat, such as the cool, moist burrows of desert rodents. Migration is a second kind of behavioral response. The most obvious kind of mammalian migration is latitudinal. Many temperate-zone bats, for example, undertake extensive migrations, although other bat species hibernate near their summer foraging grounds in caves or other equable shelters during severe weather when insects are not available. Caribou (Rangifer tarandus), or reindeer, migrate from the tundra to the forest edge in search of a suitable winter range, and a number of cetaceans (whales, dolphins, and porpoises) and pinnipeds (walruses and seals) undertake long migrations from polar waters to more temperate latitudes. Gray whales, for example, migrate southward to calving grounds along the coasts of South Korea and Baja California from summer feeding grounds in the northern Pacific Ocean (Okhotsk, Bering, and Chukchi seas). Of comparable extent is the dispersive feeding migration of the northern fur seal (Callorhinus ursinus).

Migrations of lesser extent include the elevational movements from mountains to valleys of some ungulates—the American elk (Cervus elaphus canadensis), or wapiti, and bighorn sheep (Ovis canadensis), for example—and the local migrations of certain bats from summer roosts to hibernation sites. Most migratory patterns of mammals are part of a recurrent annual cycle, but the irruptive (sudden) emigrations of lemmings and snowshoe hares are largely acyclic responses to population pressure on food supplies.

Populations

A population consists of individuals of three "ecological ages"—prereproductive, reproductive, and postreproductive. The structure and dynamics of a population depend, among other things, on the relative lengths of these ages, the rate of recruitment of individuals (either by birth or by immigration), and the rate of emigration or death. The reproductive potential of some rodents is well known; some mice are reproductively mature at four weeks of age, have gestation periods of three weeks or less, and may experience postpartum estrus, with the result that pregnancy and lactation may overlap. Litter size, moreover, may average four or more, and breeding may occur throughout the year in favourable localities. The reproductive potential of a species is, of course, a theoretical maximum that is rarely met, inasmuch as, among other reasons, a given female typically does not reproduce throughout the year. Growth of a population depends on the survival of individuals to reproductive age. The absolute age at sexual maturity ranges from less than 4 weeks in some rodents to some 15 years in the African bush elephant (Loxodonta africana).

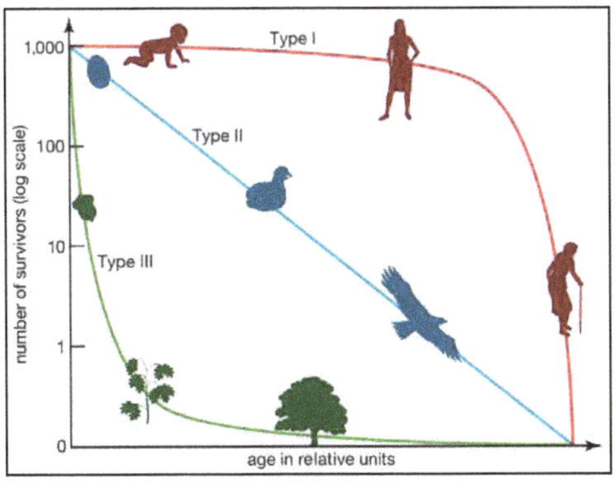

Survivorship curve: Type I, II, and III survivorship curves. A survivorship curve is the graphic representation of the number of individuals in a population that can be expected to survive to any specific age.

Post reproductive individuals are rare in most mammalian populations. Survival through more than a single reproductive season is probably uncommon in many small mammals, such as mice and shrews. Larger species typically have longer life spans than do smaller kinds, but some bats are known, on the basis of banding records, to live nearly 20 years. Many species show greater longevity in captivity than in the wild. Captive echidnas are reported to have lived more than 50 years. Horses have been reported to live more than 60 years, and elephants have lived to more than 80. Various cetaceans survive to more than 90 years of age, and research involving the dating of harpoons embedded in some Greenland right whales (Balaena mysticetus), or bowheads, suggests that Greenland right whales can live 200 years or more.

Locomotion

Specialization in habitat preference has been accompanied by locomotor adaptations. Terrestrial mammals have a number of modes of progression. The primitive mammalian stock walked plantigrade—that is, with the digits, bones of the midfoot, and parts of the ankle and wrist in contact with the ground. The limbs of ambulatory mammals are typically mobile, capable of considerable rotation.

Brown bear (Ursus arctos) running across the water in Katmai National Park and Preserve, Alaska.

Mammals modified for running are termed cursorial. The stance of cursorial species may be digitigrade (the complete digits contacting the ground, as in dogs) or unguligrade (only tips of digits contacting the ground, as in horses). In advanced groups limb movement is forward and backward in a single plane.

Saltatory (leaping) locomotion, sometimes called "ricochetal," has arisen in several unrelated groups (some marsupials, lagomorphs, and several independent lineages of rodents). This mode of locomotion is typically found in mammals living in open habitats. Jumping mammals typically have elongate, plantigrade hind feet, reduced forelimbs, and long tails. Convergent evolution within a given adaptive mode has contributed to the ecological similarity of regional mammalian faunas.

Mammals of several orders have attained great size (elephants, hippopotamuses, and rhinoceroses) and have converged on specializations for a ponderous mode of locomotion referred to as "graviportal." These animals have no digit reduction and deploy the digits in a circle around the axis of the limb for maximum support, like the pedestal of a column.

Bats are the only truly flying mammals. Only with active flight have the resources of the aerial habitat been successfully exploited. Mammals belonging to other groups (colugos, marsupials, rodents) are adapted for gliding. A gliding habit is frequently accompanied by scansorial (climbing) locomotion. Many nongliders, such as tree squirrels, are also scansorial.

Well-adapted arboreal mammals frequently are plantigrade, five-toed, and equipped with highly mobile limbs. Some species, including many New World monkeys, have a prehensile tail, which is used like a fifth hand. Brachiation, or "arm walking," in which the animal hangs from branches and moves by a series of long swings, is an adaptation seen in gibbons. The primitive opposable anthropoid thumb is reduced as a specialization for this method of locomotion. Tarsiers are highly arboreal primates that have expanded pads on the digits to improve grasping, whereas many other arboreal mammals have claws or well-developed nails.

Several mammalian groups (sirenians, cetaceans, and pinnipeds) have independently assumed fully aquatic habits. In some cases semiaquatic mammals are relatively unmodified representatives of otherwise terrestrial groups (otters, muskrats, and water shrews, for example). Other kinds have undergone profound modification for natatorial (swimming) locomotion for life at sea. Pinniped carnivores (walruses and seals) give birth to their young on land, but cetaceans are completely helpless out of water, on which they depend for mechanical support and thermal insulation.

Food Habits

The earliest mammals, like their reptilian ancestors, were active predators. From such a basal stock there has been a complex diversification (radiation) of trophic level adaptations. Modern mammals occupy a wide spectrum of feeding niches. In most terrestrial and some aquatic communities, carnivorous mammals are the top predators. Herbaceous mammals often serve as primary consumers in most ecosystems. The voracious shrews, smallest of mammals, sometimes prey on vertebrates larger than themselves. They may eat twice their weight in food each day to maintain their active metabolism and compensate for heat loss caused by an unfavourable surface-to-volume ratio. On the other hand, the largest of vertebrates, the blue whale (Balaenoptera musculus), feeds on minute planktonic crustaceans called krill.

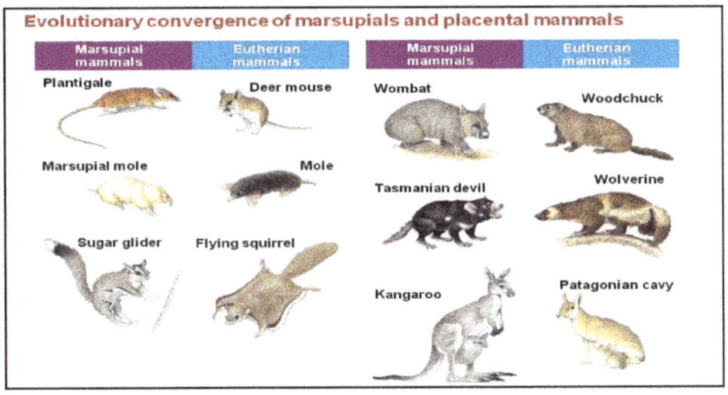

Parallel evolution of marsupial mammals in Australia and placental mammals on other continents.

Within a given lineage, the adaptive radiation of food habits may be broad. Some of the carnivores have become omnivorous (raccoons, bears) or herbivorous (giant panda). Marsupials exhibit a great variety of feeding types, and in Australia marsupials have radiated to fill ecological niches highly analogous to those of placental mammals elsewhere; there are marsupial "moles," "anteaters," "mice," "rats," "cats," and "wolves." Some bandicoots have ecological roles similar to those of rabbits, and wombats are partially burrowing (semifossorial) herbivores analogous to marmots. In Australia the niche of large grazers and browsers is filled by a variety of kangaroos and wallabies.

Within the bats there has also been a remarkable adaptive radiation of food habits. Early in the history of the order, there evidently was a divergence into insectivorous (insect-eating) and frugivorous (fruit-eating) lines. The flying foxes (Megachiroptera) have generally maintained a fruit-eating habit, although some have become rather specialized nectar feeders. Members of the other major group (Microchiroptera) have been less conservative and have undergone considerable divergence in feeding habits. A majority of living microchiropterans are insectivorous, but members of two different families have become fish eaters. Within the large Neotropical family Phyllostomatidae, there are groups specialized to feed on fruit, nectar, insects, and small vertebrates (including other bats). Aberrant members of the family are the vampire bats, with a specialized dentition to aid blood lapping.

Form and Function

Skin and Hair

The skin of mammals is constructed of two layers, a superficial nonvascular epidermis and an inner layer, the dermis, or corium. The two layers interlock via fingerlike projections (dermal papillae), consisting of sensitive vascular dermis projecting into the epidermis. The outermost layers of the epidermis are cornified (impregnated with various tough proteins), and their cells are enucleate (lacking cell nuclei). The epidermis is composed of flattened cells in layers and is the interface between the individual and the environment. Its primary function is defensive, and it is cornified to resist abrasion. The surface of the skin is coated with lipids and organic salts, the so-called "acid mantle," which is thought to possess antifungal and antibacterial properties. Deep in the epidermis is an electronegative (electron-attracting) layer, a further deterrent to foreign organic or ionic agents.

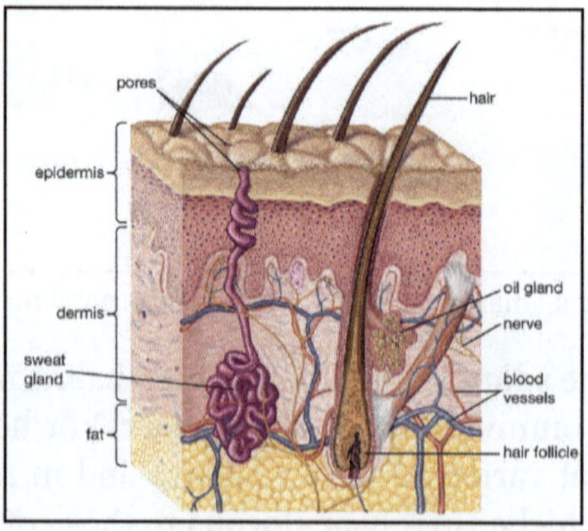

Skin: A cross section of mammalian skin and its underlying structures.

The dermis lies beneath the epidermis and nourishes it. The circulation of the dermis is variously developed in mammals, but it is typically extensive, out of proportion to the nutritional needs of the tissue. Its major role is to moderate body temperature and blood pressure by forming a peripheral shunt, an alternate route for the blood. Also in the dermis are sensory nerve endings to alert the individual to pressure (touch), heat, cold, and pain. In general, skin bearing hairs has few or no specialized sensory endings. The sensation of touch on hairy skin in humans depends on stimulation of the nerve fibres associated with the hairs. However, hairless skin, such as the lips and fingertips, has specialized endings. Hair is derived from an invagination (pocketing) of the epidermis termed a follicle. Collectively, the hair is called the pelage. The individual hair is a rod of keratinized cells that may be cylindrical or more or less flattened. Keratin is a protein also found in claws and nails. The inner medulla of the hair is hollow and contains air; in the outer cortex layer there are frequently pigment granules. Associated with the hair follicle are nerve endings and a muscle, the arrector pili. The latter allows the erection of individual hairs to alter the insulative qualities of the pelage. The follicle also gives rise to sebaceous glands that produce sebum, a substance that lubricates the hair.

Most mammals have three distinct kinds of hairs. Guard hairs protect the rest of the pelage from abrasion and frequently from moisture, and they usually lend a characteristic colour pattern. The thicker underfur is primarily insulative and may differ in colour from the guard hairs. The third common hair type is the vibrissa, or whisker, a stiff, typically elongate hair that functions in tactile sensation. Hairs may be further modified to form rigid quills. The "horn" of the rhinoceros is composed of a fibrous keratin material derived from hair. Examples of keratinized derivatives of the integument other than hair are horns, hooves, nails, claws, and baleen.

Even though the primary function of the skin is defensive, it has been modified in mammals to serve such diverse functions as thermoregulation and nourishment of young. Secretions of sweat glands promote cooling due to evaporation at the surface of the body, and mammary glands are a type of apocrine gland.

In certain groups (primates in particular) the skin of the face is under intricate muscular control, and movements of the skin express and communicate emotion. In many mammals the colour and pattern of the pelage are important in communicative behaviour. Patterns may be startling (dymantic), as seen in the mane of the male lion or hamadryas baboon, warning (sematic), as seen in the bold pattern of skunks, or concealing (cryptic), perhaps the most common adaptation of pelage colour.

Hair has been secondarily lost or considerably reduced in some kinds of mammals. In adult cetaceans insulation is provided by thick subcutaneous fat deposits, or blubber, with hair limited to a few stiff vibrissae about the mouth. The bare skin is one of a number of features that contribute to the remarkably advanced hydrodynamics of locomotion in the group. Some burrowing (fossorial) mammals also tend toward reduction of the hair. This is shown most strikingly by the sand rats of northeastern Africa, but considerable loss of hair has also occurred in some species of pocket gophers. Hair may also be lost on restricted areas of the skin, as from the face in many monkeys or the buttocks of mandrills, and may be sparse on elephants and such highly modified species as pangolins and armadillos.

Continuous growth of hair (indeterminate), as seen on the heads of humans, is rare among mammals. Hairs with determinate growth are subject to wear and must be replaced periodically—a process termed molt. The first coat of a young mammal is referred to as the juvenal pelage, which typically is of fine texture like the underfur of adults and is replaced by a postjuvenile molt. Juvenal pelage is succeeded either directly by adult pelage or by the subadult pelage, which in some species is not markedly distinct from that of the adult. Once this pelage has been acquired, molting continues to recur at intervals, often annually or semiannually and sometimes more frequently. The pattern of molt typically is orderly, but it varies widely between species. Some mammals apparently molt continuously, with a few hairs at a time replaced throughout the year.

Teeth

Specialization in food habits has led to profound dental changes. The primitive mammalian tooth had high, sharp cusps and served to tear flesh or crush chitinous material (primarily the exoskeletons of terrestrial arthropods, such as insects). Herbivores tend to have specialized cheek teeth with complex patterns of contact (occlusion) and various ways of expanding the crowns of the teeth and circumventing the problem of wear. Omnivorous mammals, such as bears, pigs, and humans, tend to have molars with low, rounded cusps, termed bunodont.

A yawning African lion (Panthera leo) showing its long canine teeth.

A prime example of convergence in conjunction with dietary specialization is seen in those mammals adapted to feeding on ants and termites, a specialization generally termed myrmecophagy ("ant eating"). Trends frequently associated with myrmecophagy include strong claws, an elongate rounded skull, a wormlike extensible tongue, marked reduction in the mandible (lower jaw), and loss or extreme simplification of the teeth (dentition). This habit has led to remarkably similar morphology among animals as diverse as the echidna (a monotreme), the numbat (a marsupial), the anteater (a xenarthran), the aardvark (a tubulidentate), and the pangolin (a pholidotan).

Specialized herbivores evolved early in mammalian history. The extinct multituberculates were the earliest mammalian herbivores and have the longest evolutionary history, lasting more than 100 million years from 178 million to 50 million years ago. Multituberculate fossils, such as those of Ptilodus, dated to the Paleocene Epoch (66–56 million years ago) of North America, have been found on all continents. Similarities in teeth not due to common ancestry have occurred widely in herbivorous groups. Most herbivores have incisors modified for nipping or gnawing, have lost teeth with the resultant development of a gap (diastema) in the tooth row, and exhibit some molarization (expansion and flattening) of premolars to expand the grinding surface of the cheek teeth. Rootless incisors or cheek teeth have evolved frequently, their open pulp cavity allowing continual growth throughout life. Herbivorous specializations have evolved independently in multituberculates, rodents, lagomorphs, primates, and the wide diversity of ungulate and subungulate orders.

Skeleton

The mammalian skeletal system shows a number of advances over that of lower vertebrates. The mode of ossification (bone formation) of the long bones is characteristic.

In lower vertebrates each long bone has a single centre of ossification (the diaphysis), and replacement of cartilage by bone proceeds from the centre toward the ends, which may remain cartilaginous, even in adults. In mammals secondary centres of ossification (the epiphyses) develop at the ends of the bones. Growth of bones occurs in zones of cartilage between diaphysis and epiphyses. Mammalian skeletal growth is termed determinate, for once the actively growing zone of cartilage has been obliterated, growth in length ceases. As in all bony vertebrates, of course, there is continual renewal of bone throughout life. The advantage of epiphyseal ossification lies in the fact that the bones have strong articular (joint-related) surfaces before the skeleton is mature. In general, the skeleton of the adult mammal has less structural cartilage than does that of a reptile.

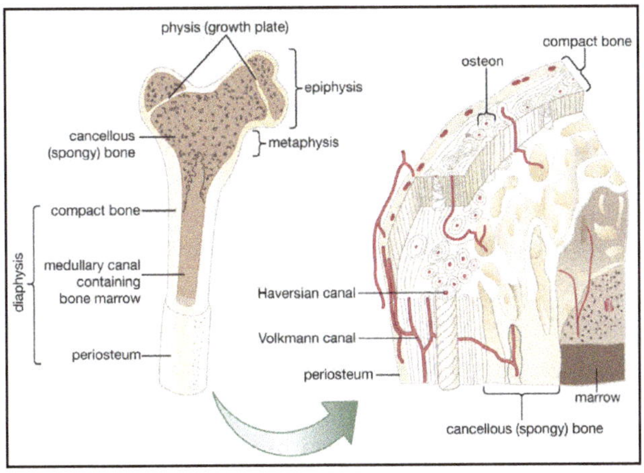

Internal structure of a human long bone, with a magnified cross section of the interior. The central tubular region of the bone, called the diaphysis, flares outward near the end to form the metaphysis, which contains a largely cancellous, or spongy, interior. At the end of the bone is the epiphysis, which in young people is separated from the metaphysis by the physis, or growth plate. The periosteum is a connective sheath covering the outer surface of the bone. The Haversian system, consisting of inorganic substances arranged in concentric rings around the Haversian canals, provides compact bone with structural support and allows for metabolism of bone cells. Osteocytes (mature bone cells) are found in tiny cavities between the concentric rings. The canals contain capillaries that bring in oxygen and nutrients and remove wastes. Transverse branches are known as Volkmann canals. The skeletal system of mammals and other vertebrates is broadly divisible functionally into axial and appendicular portions. The axial skeleton consists of the braincase (cranium) and the backbone and ribs, and it serves primarily to protect the central nervous system. The limbs and their girdles constitute the appendicular skeleton. In addition, there are skeletal elements derived from the gill arches of primitive vertebrates, col-lectively termed the visceral skeleton. Visceral elements in the mammalian skeleton include the jaws, the hyoid apparatus supporting the tongue, and the auditory ossicles of the middle ear. The postcranial axial skeleton in mammals generally has remained rather conservative during the course of evolution.

The vast majority of mammals have seven cervical (neck) vertebrae; exceptions are sloths, with six or nine cervicals, and the sirenians with six. The anterior two cervical vertebrae are differentiated as atlas and axis. Specialized articulations of these two bones allow complex movements of the head on the trunk. Thoracic vertebrae bear ribs and are variable in number. The anterior ribs converge toward the ventral midline to articulate with the sternum, or breastbone, forming a semirigid thoracic "basket" for the protection of heart and lungs. Posterior to the thoracic region are the lumbar vertebrae, ranging from 2 to 21 in number (most frequently 4 to 7). Mammals have no lumbar ribs. There are usually 3 to 5 sacral vertebrae, but some xenarthrans have as many as 13. Sacral vertebrae fuse to form the sacrum, to which the pelvic girdle is attached. Caudal (tail) vertebrae range in number from 5 (fused elements of the human coccyx [or tailbone]) to 50.

The basic structure of the vertebral column is comparable throughout the Mammalia, although in many instances modifications have occurred in specialized locomotor modes to gain particular mechanical advantages. The vertebral column and associated muscles of many mammals are structurally analogous to a cantilever girder.

The skull is composite in origin and complex in function. Functionally the bones of the head are separable into the braincase and the jaws. In general, it is the head of the animal that meets the environment. The skull protects the brain and sense capsules (the parts of the skeleton that facilitate the senses of sight, hearing, taste, and smell), houses the teeth and tongue, and contains the entrance to the pharynx. Thus, the head functions in sensory reception, food acquisition, defense, respiration, and (in higher groups) communication. To serve these functions, bony elements have been recruited from the visceral skeleton, the endochondral skeleton (the parts of the skeleton that form from cartilage), and the dermal skeleton of lower vertebrates.

The skull of mammals differs markedly from that of reptiles because of the great expansion of the brain. The sphenoid bones that form the reptilian braincase form only the floor of the braincase in mammals. The side is formed in part by the alisphenoid bone, derived from the epipterygoid, a part of the reptilian palate. Dermal elements, the frontals and parietals, have come to lie deep to (beneath) the muscles of the jaw to form the dorsum of the braincase. Reptilian dermal roofing bones, lying superficial to the muscles of the jaw, are represented in mammals only by the jugal bone of the zygomatic arch, which lies under the eye.

In mammals a secondary palate is formed by processes of the maxillary bones and the palatines, with the pterygoid bones reduced in importance. The secondary palate separates the nasal passages from the oral cavity and allows continuous breathing while chewing or suckling.

Other specializations of the mammalian skull include paired articulating surfaces at the neck (occipital condyles) and an expanded nasal chamber with complexly folded turbinal bones, providing a large area for detection of odours. Eutherians have evolved bony protection for the middle ear, the auditory bulla. The development of this structure varies, although a ring-shaped (annular) tympanic bone is always present.

The bones of the mammalian middle ear are a diagnostic feature of the class. The three auditory ossicles form a series of levers that serve mechanically to increase the amplitude of sound waves reaching the tympanic membrane, or eardrum, produced as disturbances of the air. The innermost bone is the stapes, or "stirrup bone." It rests against the oval window of the inner ear. The stapes is homologous with the entire stapedial structure of reptiles, which in turn was derived from the hyomandibular arch of primitive vertebrates. The incus, or "anvil," articulates with the stapes. The incus was derived from the quadrate bone, which is involved in the jaw articulation in reptiles. The malleus, or "hammer," rests against the tympanic membrane and articulates with the incus. The malleus is the homologue of the reptilian articular bone. The mechanical efficiency of the middle ear has thus been increased by the incorporation of two bones of the reptilian jaw assemblage. In mammals the lower jaw is a single bone, the dentary, which articulates with the squamosal of the skull.

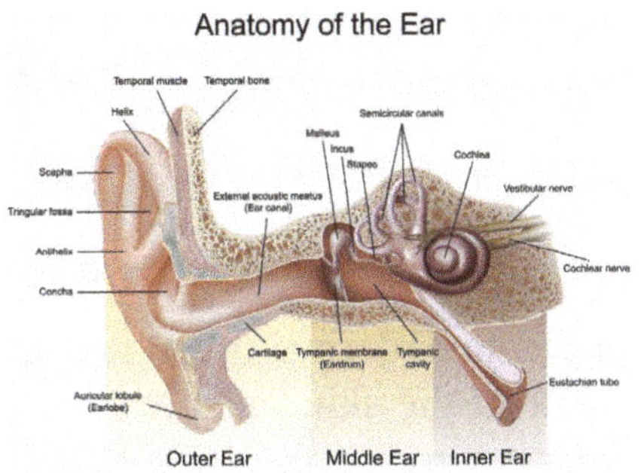

The structures of the outer, middle, and inner ear.

The limbs and girdles have been greatly modified with locomotor adaptations. The ancestral mammal had well-developed limbs and was five-toed. In each limb there were two distal (outer) elements (radius and ulna in the forelimb; tibia and fibula in the hind limb) and a single proximal (inner or upper) element (humerus; femur). There were nine bones in the wrist, the carpals, and seven bones in the ankle, the tarsals. The phalangeal formula (the number of phalangeal bones in each digit, numbered from inside outward) is 2-3-3-3-3 in primitive mammals; in primitive reptiles it is 2-3-4-5-3. Modifications in mammalian limbs have involved reduction, loss, or fusion of bones. Loss of the clavicle from the shoulder girdle, reduction in the number of toes, and modifications of tarsal and carpal bones are typical correlates of cursorial locomotion. Scansorial and arboreal groups tend to maintain or emphasize the primitive divergence of the thumb and hallux (the inner toe on the hind foot).

Centres of ossification sometimes develop in nonbony connective tissue. Such bones are termed heterotopic or sesamoid elements. The kneecap (patella) is such a bone. Another important bone of this sort, found in many kinds of mammals, is the baculum, or os penis, which occurs as a stiffening rod in the penis of such groups as carnivores, many bats, rodents, some insectivores, and many primates. The os clitoridis is a homologous structure found in females.

Muscles

The muscular system of mammals is generally comparable to that of reptiles. With changes in locomotion, the proportions and specific functions of muscular elements have been altered, but the relationships of these muscles remain essentially the same. Exceptions to this generalization are the muscles of the skin and of the jaw.

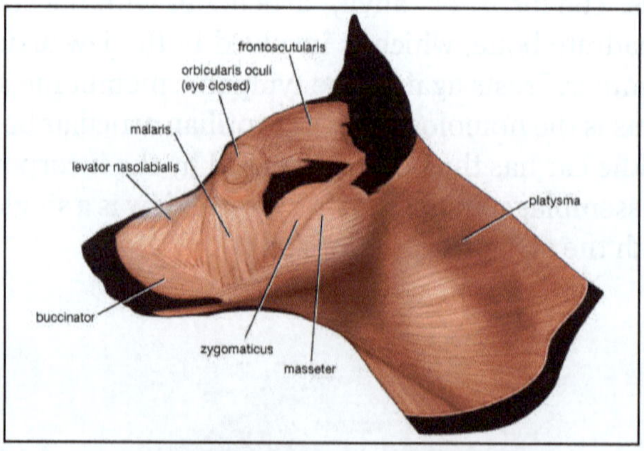

Lateral view of facial musculature of a dog.

The panniculus carnosus is a sheath of dermal (skin) muscle, developed in many mammals that allows the movement of the skin independent of the movement of deeper muscle masses. These movements function in such mundane activities as the twitching of the skin to foil insect pests and in some species also are important in shivering, a characteristic heat-producing response to thermal stress. The dermal musculature of the facial region is particularly well developed in primates and carnivores but occurs in other groups as well. Facial mobility allows expression that may be of importance in the behavioral maintenance of interspecific social structure. The temporalis muscle is the major adductor (closer) of the reptilian jaw. In mammals the temporalis is divided into a deep temporalis proper and a more superficial masseter muscle. The temporalis attaches to the coronoid process of the mandible (lower jaw) and the temporal bone of the skull. The masseter passes from the angular process of the mandible to the zygomatic arch. The masseter allows an anteroposterior (forward-backward) movement of the jaw and is highly developed in mammals, such as rodents, for which grinding is the important function of the dentition.

Digestive System

The alimentary canal is highly specialized in many kinds of mammals. In general, specializations of the gut accompany herbivorous habits. The intestines of herbivores are typically elongate, and the stomach may also be specialized. Subdivision of the gut allows areas of differing physiological environments for the activities of different sorts of enzymes and symbiotic bacteria, which aid the animal by breaking down certain compounds that are otherwise undigestible. In ruminant artiodactyls, such as antelopes, deer, and cattle, the stomach has up to four chambers, each with a particular function in the processing of vegetable material. A cecum is common in many herbivores. The cecum is a blind sac at the far end of the small intestine where complex compounds such as cellulose are acted upon by symbiotic bacteria. The vermiform appendix is a diverticulum of the cecum. The appendix is rich in lymphoid tissue and in many mammals is concerned with defense against toxic bacterial products.

Hares and rabbits, the sewellel, or "mountain beaver" (Aplodontia rufa), and some insectivores exhibit a phenomenon of reingestion called coprophagy, in which at intervals specialized fecal pellets are produced. These pellets are eaten and passed through the alimentary canal a second time. Where known to be present, this pattern seems to be obligatory. Reingestion primarily occurs in members of the shrew, rodent, and rabbit groups; however, the behaviour has been observed to a lesser degree in other groups, including canines and pikas. The process appears to allow the animal to absorb in the upper gut vitamins produced by the microflora of the lower gut but not absorbable there.

Excretory System

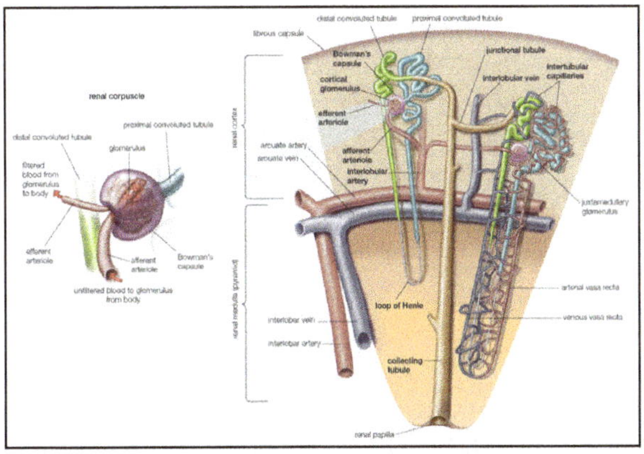

Each nephron of the kidney contains blood vessels and a special tubule. As the filtrate flows through the tubule of the nephron, it becomes increasingly concentrated into urine.

The mammalian kidney is constructed of a large number of functional units called nephrons. Each nephron consists of a distal tubule, a medial section termed the loop of Henle, a proximal tubule, and a renal corpuscle. The renal corpuscle is a knot of capillaries (glomerulus) surrounded by a sheath (Bowman's capsule). The renal corpuscle is a pressure filter, relying on blood pressure to remove water, ions, and small organic molecules from the blood. Some of the material removed is waste, but some is of value to the organism. The filtrate is sorted by the tubules, and water and needed solutes are resorbed. Resorption is both passive (osmotic) and active (based on ion transport systems). The distal convoluted tubules drain into collecting tubules, which in turn empty into the calyces, or branches, of the renal pelvis, the expanded end of the ureter.

The pressure-pump nephron of mammals is so efficient that the renal portal system of lower vertebrates has been completely lost. Mammalian kidneys show considerable variety in structure, relative to the environmental demands on a given species. In particular, desert rodents have long loops of Henle and are able to resorb much water and to excrete a highly concentrated urine. Urea is the end product of protein metabolism in mammals, and excretion is therefore called ureotelic.

Reproductive System

The Male Tract

The testes of mammals descend from the abdominal cavity to lie in a compartmented pouch termed the scrotum. In some species the testes are permanently scrotal, and the scrotum is sealed off from the general abdominal cavity. In other species the testes migrate to the scrotum only during the breeding season. It is thought that the temperature of the abdominal cavity is too high to allow spermatogenesis; the scrotum allows cooling of the testes.

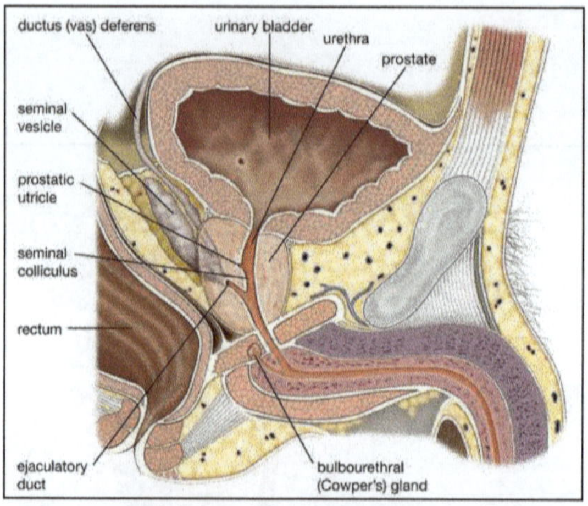

The transport of spermatozoa is comparable to that in reptiles, relying on ducts derived from urinary ducts of earlier vertebrates. Mammalian specialities are the bulbourethral (or Cowper's) glands, the prostate gland, and the seminal vesicle or vesicular gland. Each of these glands adds secretions to the spermatozoa to form semen, which passes from the body via a canal (urethra) in the highly vascular, erectile penis. The tip of the penis, the glans, may have a complex morphology and has been used as a taxonomic character in some groups. The penis may be retracted into a sheath along the abdomen or may be pendulous, as in bats and many primates.

The Female Tract

The structure of the female reproductive tract is variable. Four types of uterus are generally recognized among placentals, based on the relationship of the uterine horns (branches). A duplex uterus characterizes rodents and rabbits; the uterine horns are completely separated and have separate cervices opening into the vagina. Carnivores have a bipartite uterus, in which the horns are largely separate but enter the vagina by a single cervix. In the bicornate uterus, typical of many ungulates, the horns are distinct for less than half their length; the lower part of the uterus is a common chamber, the body. Higher primates have a simplex uterus in which there is no separation between the horns and thus a single chamber.

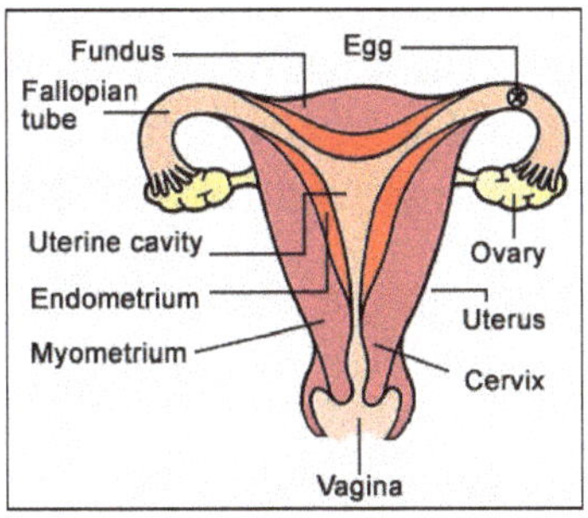

The uterus is an inverted pear-shaped muscular organ of the female reproductive system, located between the bladder and the rectum. It functions to nourish and house the fertilized egg until the unborn child is ready to be delivered.

The female reproductive tract of marsupials is termed didelphous; the vagina is paired, as are oviducts and uteri. In primitive marsupials there are paired vaginae lateral to the ureters. In more advanced groups, such as kangaroos, the lateral vaginae persist and conduct the migration of spermatozoa, but a medial "pseudovagina" functions as the birth canal.

Monotremes have paired uteri and oviducts, which empty into a urogenital sinus (cavity) as fluid wastes do. The sinus passes into the cloaca, a common receptacle for reproductive and excretory products.

Circulatory System

In mammals, as in birds, the right and left ventricles of the heart are completely separated, so that pulmonary (lung) and systemic (body) circulations are completely independent. Oxygenated blood arrives in the left atrium from the lungs and passes to the left ventricle, whence it is forced through the aorta to the systemic circulation. Deoxygenated blood from the tissues returns to the right atrium via a large vein, the vena cava, and is pumped to the pulmonary capillary bed through the pulmonary artery.

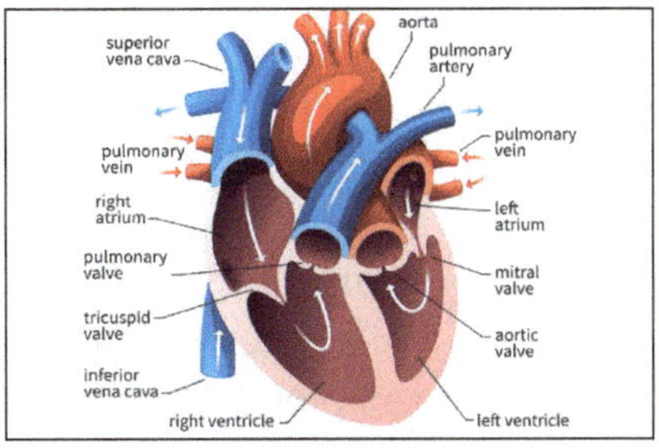

Cross section of a four-chambered mammalian heart.

Among vertebrates contraction of the heart is myogenic, or generated by muscle; rhythm is inherent in all cardiac muscle, but in myogenic hearts the pacemaker is derived from cardiac tissue. The pacemaker in mammals (and also in birds) is an oblong mass of specialized cells called the sinoatrial node, located in the right atrium near the junction with the venae cavae. A wave of excitation spreads from this node to the atrioventricular node, which is located in the right atrium near the base of the interatrial septum. From this point excitation is conducted along the atrioventricular bundle (bundle of His) and enters the main mass of cardiac tissue along fine branches, the Purkinje fibres. Homeostatic, or stable, control of the heart by neuroendocrine or other agents is mediated through the intrinsic control network of the heart.

Blood leaves the left ventricle through the aorta. The mammalian aorta is an unpaired structure derived from the left fourth aortic arch of the primitive vertebrate. Birds, on the other hand, retain the right fourth arch.

The circulatory system forms a complex communication and distribution network to all physiologically active tissues of the body. A constant, copious supply of oxygen is required for sustaining the active, heat-producing (endothermous) physiology of the higher vertebrates. The efficiency of the four-chambered heart is important to this function. Oxygen is transported by specialized red blood cells, or erythrocytes, as in all vertebrates. Packaging the oxygen-bearing pigment hemoglobin in erythrocytes keeps the viscosity of the blood minimal and thereby allows efficient circulation while limiting the mechanical load on the heart. The mammalian erythrocyte is a highly evolved structure; its discoid, biconcave shape allows maximal surface area per unit volume. When mature and functional, mammalian red blood cells are enucleate (lacking a nucleus).

Respiratory System

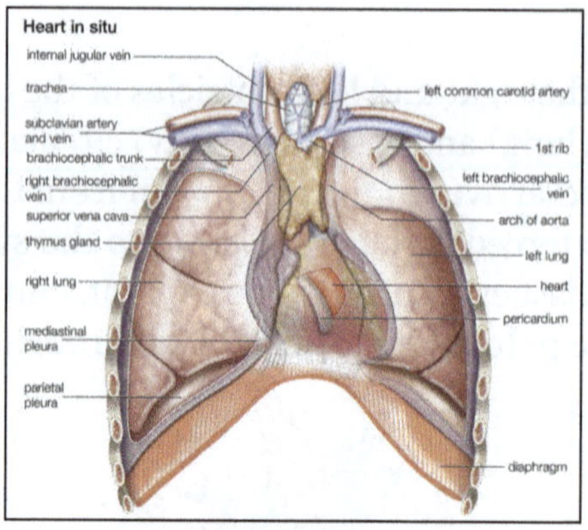

The human heart in situ.

Closely coupled with the circulatory system is the ventilatory (breathing) apparatus, the lungs and associated structures. Ventilation in mammals is unique. The lungs themselves are less efficient than those of birds, for air movement consists of ebb and flow, rather than a one-way circuit, so a residual volume of air always remains that cannot be expired. Ventilation in mammals is by means of a negative pressure pump made possible by the evolution of a definitive thoracic cavity with a diaphragm.

The diaphragm is a unique composite structure consisting of (1) the transverse septum (a wall that primitively separates the heart from the general viscera); (2) pleuroperitoneal folds from the body wall; (3) mesenteric folds; and (4) axial muscles inserting on a central tendon, or diaphragmatic aponeurosis.

The lungs lie in separate airtight compartments called pleural cavities, separated by the mediastinum. As the size of the pleural cavity is increased, the lung is expanded and air flows in passively. Enlargement of the pleural cavity is produced by contraction of the diaphragm or by elevation of the ribs. The relaxed diaphragm domes upward, but when contracted it stretches flat. Expiration is an active movement brought about by contraction of abdominal muscles against the viscera.

Air typically enters the respiratory passages through the nostrils, where it may be warmed and moistened. It passes above the bony palate and the soft palate and enters the pharynx. In the pharynx the passages for air and food cross. Air enters the trachea, which divides at the level of the lungs into primary bronchi. A characteristic feature of the trachea of many mammals is the larynx. Vocal cords stretch across the larynx and are vibrated by forced expiration to produce sound. The laryngeal apparatus may be greatly modified for the production of complex vocalizations. In some groups—for example, howler monkeys—the hyoid apparatus is incorporated into the sound-producing organ, as a resonating chamber.

Nervous and Endocrine Systems

The nervous system and the endocrine system are closely related to one another in their function, for both serve to coordinate activity. The endocrine glands of mammals generally have more complex regulatory functions than do those of lower vertebrates. This is particularly true of the pituitary gland, which supplies hormones that regulate the reproductive cycle. Follicle-stimulating hormone (FSH) initiates the maturation of the ovarian follicle. Luteinizing hormone (LH) mediates the formation of the corpus luteum from the follicle following ovulation. Prolactin, also a product of the anterior pituitary, stimulates the secretion of milk.

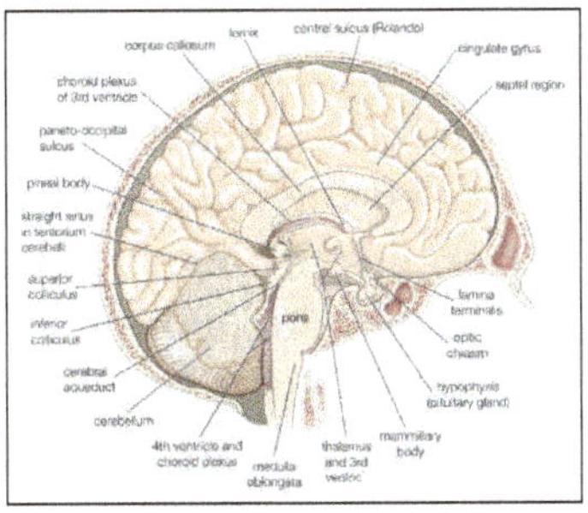

Medial view of the left hemisphere of the human brain.

Control of the pituitary glands is partially by means of neurohumours from the hypothalamus, a part of the forebrain in contact with the pituitary gland by nervous and circulatory pathways. The hypothalamus is of the utmost importance in mammals, for it integrates stimuli from both internal and external environments, channeling signals to higher centres or into autonomic pathways.

The cerebellum of vertebrates is at the anterior end of the hindbrain. Its function is to coordinate motor activities and to maintain posture. In most mammals the cerebellum is highly developed, and its surface may be convoluted to increase its area. The data with which the cerebellum works arrive from proprioceptors ("self-sensors") in the muscles and from the membranous labyrinth of the inner ear, the latter giving information on position and movements of the head.

In the vertebrate ancestors of mammals, the cerebral hemispheres were centres for the reception of olfactory stimuli. Vertebrate evolution has favoured an increasing importance of these lobes in the integration of stimuli. Their great development in mammals as centres of association is responsible for the "creative" behaviour of members of the class—i.e., the ability to learn, to adapt as individuals to short-term environmental change through appropriate responses on the basis of previous experience. In vertebrate evolution the gray matter of the cerebrum has moved from a primitive internal position in the hemispheres to a superficial position. The superficial gray matter is termed the pallium. The paleopallium of amphibians has become the olfactory lobes of the higher vertebrates; the dorsolateral surface, or archipallium, has become the mammalian hippocampus. The great neural advance of the mammals lies in the elaboration of the neopallium, which makes up the bulk of the cerebrum. The neopallium is an association centre, the dominant centre of neural function, and is involved in so-called "intelligent" response. By contrast, the highest centre in the avian brain is the corpus striatum, an evolutionary product of the basal nuclei of the amphibian brain. Therefore, the bulk of complex behaviour of birds is instinctive. The surface of the neopallium tends in some mammals to be greatly expanded by convoluting, forming folds (gyri) between deep grooves (sulci).

Evolution and Classification

The Evolution of Mammalian Condition

Mammals were derived in the Triassic Period (about 252 million to 201 million years ago) from members of the reptilian order Therapsida. The therapsids, members of the subclass Synapsida (sometimes called the mammal-like reptiles), generally were unimpressive in relation to other reptiles of their time. Synapsids were present in the Carboniferous Period (about 359 million to 299 million years ago) and are one of the earliest known reptilian groups. They were the dominant reptiles of the Permian Period (299 million to 252 million years ago), and, although they were primarily predaceous in habit, the adaptive radiation included herbivorous species as well. In the Mesozoic

Era (about 252 million to 66 million years ago), the most important of the synapsids were the archosaurs, or “ruling reptiles,” and the therapsids were, in general, small active carnivores. Therapsids tended to evolve a specialized heterodont dentition (that is, a set of teeth separated into molars, incisors, and canines) and to improve the mechanics of locomotion by bringing the plane of action of the limbs close to the trunk. A secondary palate was developed, and the temporal musculature, the muscle involved in closing the jaw, was expanded.

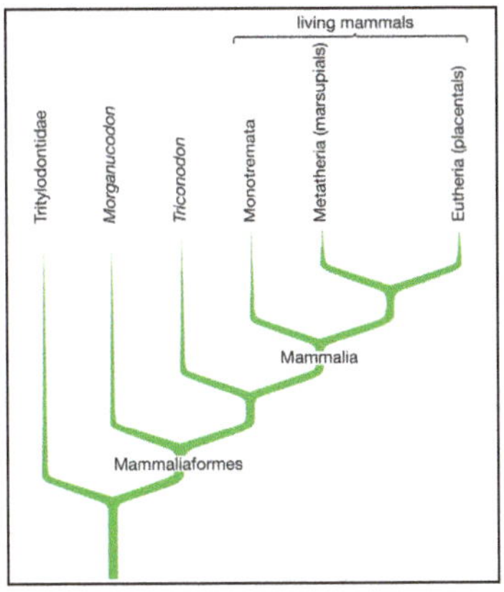

Mammal phylogeny.

The several features that separate modern reptiles from modern mammals doubtless evolved at different rates. Many attributes of mammals are correlated with their highly active habit—for example, efficient double circulation with a completely four-chambered heart, anucleate and biconcave erythrocytes, the diaphragm, and the secondary palate (which separates passages for food and air and allows breathing during mastication or suckling). Hair for insulation is a correlate of endothermy, or warm-bloodedness, the physiological maintenance of individual temperature independent of environmental temperature. Endothermy allows high levels of sustained activity. The unique characteristics of mammals thus would seem to have evolved as a complex interrelated system.

Because the characteristics that separate reptiles and mammals evolved at different rates and in response to a variety of interrelated conditions, at any point in the period of transition from reptiles to mammals, there were forms that combined various characteristics of both groups. Such a pattern of evolution is termed mosaic and is a common phenomenon in those transitions marking the origin of major new adaptive types. To simplify definitions and to allow the strict delimitation of the Mammalia, some authors have suggested basing the boundary on a single characteristic, the articulation of the jaw between the dentary and squamosal bones and the attendant movement of accessory jawbones to the middle ear as auditory ossicles. The use of a single osteological character allows the placement in a logical classification of numerous fossil species, other mammalian characteristics of which, such as the degree of endothermy and nursing of young and the condition of the internal organs, probably never will be evaluated. It must be recognized, however, that were the advanced therapsids alive today, taxonomists would be hard put to decide which to place in the Reptilia and which in the Mammalia.

Morganucodon: The extinct Morganucodon lived 200 million years ago.

Prototheria

The subclass Prototheria contains the egg-laying mammals, which are the most ancestral forms in the class Mammalia. There are only three extant species grouped into two families and a single order, the Monotremata. Despite bearing fewer species than most mammalian genera, the prototherians are so unique among mammals that there is little question that they represent a distinct and ancient branch of the mammmalian family tree. However, it is not clear how monotremes are related to the two other major lineages of mammals, marsupials (Metatheria) and placentals (Eutheria). Some evidence supports the hypothesis that prototherians form a clade with the marsupials, while other evidence suggests that prototherians are sister to a clade containing both marsupials and placentals.

Prototherians probably split from the lineage leading to other mammals sometime in the Mesozoic. They retain many characters of their therapsid ancestors (for example, a complex pectoral girdle, laying of eggs rather than bearing live young, limbs oriented with humerus and femur held lateral to body, and a cloaca). The skulls of monotremes are almost birdlike in appearance, with a long rostrum and smooth external appearance. Modern monotremes lack teeth as adults; sutures are hard to see; the rostrum is elongate, beak-like, and covered by a leathery sheath; and lacrimal bones are absent. Monotremes have several important mammalian characters, however, including fur (but they lack vibrissae), a four chambered heart, a single dentary bone, three middle ear bones, and the ability to lactate.

Geographic Range

Monotremes are restricted to Australia and New Guinea. Their fossil record is very poor; the earliest fossil attributed to this group is from the early Cretaceous. A fossil from Argentina suggests that the monotremes were more widely distributed early in their history.

Habitat

Prototherians are either terrestrial (Tachyglossidae) or primarily aquatic (Ornithorhynchidae). Their terrestrial habitats include deserts, sandy plains, rocky areas, and forests in both lowlands and mountains. Platypuses inhabit lakes, ponds and streams; they shelter in burrows along the banks and spend much of their time foraging in the water.

- Habitat Regions: temperate; tropical; terrestrial; freshwater.
- Terrestrial Biomes: desert or dune; savanna or grassland; forest; scrub forest; mountains.
- Aquatic Biomes: lakes and ponds; rivers and streams.
- Other Habitat Features: riparian.

Physical Description

Besides the absence of teeth, lacrimals, and obvious sutures, prototherians share a number of skeletal characteristics. On the skulls, the jugals are reduced or absent, the dentary is a slender bone with only a vestige of a coronoid process, the angle of the dentary is not inflected medially (unlike that of marsupials), auditory bullae are missing (part of the middle ear is enclosed by tympanic rings), and much of the wall of the braincase is made up by the petrosal rather than the alisphenoid (unlike all other modern mammals). Postcranially, the skeleton of prototherians is also unique among mammals. It is a fascinating mosaic of primitive characteristics inherited from therapsids but found in no other living mammals, and modifications probably related to the burrowing habits of modern prototherians. Their shoulder girdles are complex, including the standard components of modern mammals (scapula and clavicle), but also additional elements including coracoid, epicoracoid, and interclavicle. The scapula, however, is simplified, lacking a supraspinous fossa. The shoulder girdle is much more rigidly attached to the axillary skeleton than in other mammals. Femur and humerus are held roughly parallel to the ground when the animal walks, more in the fashion of therapsids and most modern reptiles than like modern mammals. Ribs are found on the neck (cervical) vertebrae as well as the chest (thoracic) vertebrae; in all other modern mammals, they are restricted to the thoracic region.

Another interesting skeletal characteristic of prototherians is the large epipubic bones in the pelvic region. Epipubic bones were originally thought to be related to having a pouch, but they are found in both males and females. They also occur in all species of marsupials, whether a pouch is present or not (not all marsupials have a pouch). It is now thought that epipubic bones are a vestige of the skeleton of therapsids, providing members of that group with extra attachments for abdominal muscles to support the weight of the hindquarters.

Prototherians are endothermic, but they have unusually low metabolic rates and maintain a body temperature that is lower than that of most other mammals.

All male prototherians have spurs on their ankles that are presumed to be used in fighting and in defense. In one family (Ornithorhynchidae), a groove along the spur carries poison secreted by adjacent glands.

- Other Physical Features: endothermic; homoeothermic; bilateral symmetry; venomous.

Reproduction

Little is known about the mating systems of prototherians. They are solitary for most of the year, coming together only to mate. During the mating season, duck-billed platypuses are found in pairs, but despite these observations, platypuses are not likely to be monogamous because males do not associate with females post-copulation, nor do they provide any parental care. Female short-nosed spiny echidnas have been observed with several males at a time, which may reflect a polygyny or polyandry. Even less can be inferred about the mating systems of long-nosed spiny echidnas because so little is known about their basic behavior and biology.

Prototherians are seasonal breeders. Typically, the breeding season lasts 1 to 3 months between July and October. At least one species (duck-billed platypuses) perform somewhat elaborate courtship behaviors prior to copulation.

The eggs layed by monotremes are small (13 to 15 mm diameter) and covered by a leathery shell. The number of eggs laid is small, usually 1 to 3, and they are placed in the mother's pouch. They contain a large yolk, which is concentrated at one end of the egg, very much like the yolk of a bird's egg. Only the left ovary is functional in the platypus, but both produce eggs in the echidna. Like the eggs of birds, monotreme eggs are incubated and hatched outside the body of the mother. Incubation lasts about 12 days. The young, which are tiny and at a very early stage of development when they hatch, break out of the eggs using a "milk tooth. They are protected in a temporary pouch in echidnas but not platypuses. They are fed milk produced by mammary glands; the milk is secreted onto the skin within the pouch and sucked or lapped up by the babies. Weaning takes place when the young are 16 to 20 weeks old.

- Key Reproductive Features: iteroparous; seasonal breeding; gonochoric/gonochoristic/dioecious (sexes separate); sexual oviparous.

The parental investment of male prototherians appears to consist entirely of acquiring mates and fertilizing a female's eggs. All other investment and parental care is provided by females. Young are born in a highly altricial state and require considerable care and protection from their mothers. As mammals, females produce milk and nurse their young. Echidnas develop a brood pouch on their abdomen within which eggs and hatched young develop for nearly two months. Young are weaned by about three months of age. Platypuses do not have a brood pouch, and instead lay their eggs in deep, complex burrows on the banks of streams and ponds. Young develop within the burrow and are weaned after 3 months.

- Parental Investment: altricial; pre-fertilization (provisioning protecting female); pre-hatching/birth (provisioning: female, protecting: female); pre-weaning/fledging (provisioning: female, protecting: female).

Lifespan/Longevity

Little is known regarding the natural lifespan of prototherians; however, they can live several decades in captivity. In at least one case, a short-nosed echidna lived 50 years.

Behavior

Prototherians are primarily solitary animals and at least one species (Tachyglossus aculeatus) is territorial. Activity patterns vary among species, and even among populations of T. aculeatus; prototherians may be diurnal, crepuscular or nocturnal. Echidnas are fully terrestrial and eat mainly ants, termites, and worms whereas platypuses spend much of their time foraging in the water for a wider variety of invertebrates. All three species are exceptional diggers, using powerful limbs to dig shelters or to quickly escape from predators. In addition to digging their way out of trouble, echidnas can roll up and erect their spines as a defense mechanism. If insufficient food is available, prototherians may enter temporary torpor or more prolonged periods of hibernation when food is scarce in the winter.

- Key Behaviors : terricolous; fossorial; natatorial; diurnal; nocturnal; crepuscular; motile; hibernation; solitary; territorial.

Communication and Perception

Hearing, olfaction, touch, and vision are all important to some degree in prototherians. Hearing and sight are well developed in platypuses and moderately well-developed in echidnas. The sense of touch is perhaps most important to a platypus that is searching at the bottom of a stream for food or an echida that is rooting through the earth for termites or worms. Platypus bills and echidna snouts are extremely sensitive organs that are essential to effective foraging. Platypuses may even use electrical stimuli to locate prey. Olfaction is well-developed in echidnas and may be used in individual recognition. Prototherians occasionally produce some simple vocalizations, but their function is unknown.

- Communication Channels: visual; tactile; acoustic; chemical.
- Perception Channels: visual; tactile; acoustic; chemical.

Food Habits

All prototherians are carnivorous, with their diets consisting of various invertebrates. Platypuses forage in the benthos of lakes and streams, using their sensitive bills to find prey. They are generalist predators, whereas echidnas specialize on either ants and termites (Tachyglossus) or worms (Zaglossus). Both species of echidna are powerful diggers and use their claws and snouts to root through the earth to find food.

- Primary Diet: carnivore (insectivore, eats non-insect arthropods, molluscivore).

Predation

Little is known about the predators of prototherians or how predation impacts prototherian populations. With their robust spines, echidnas are certainly well-protected from threats. To deter potential predators, echidnas erect their spines, roll into protected balls, or rapidly dig a hole or enter a crevice, exposing only their spines.

Ecosystem Roles

Prototherians may significantly impact populations of their prey; this may be more true for echidnas because they specialize on only a few prey types rather than eating a little bit of many different species. Because they are adept diggers, prototherians create and modify habitat for other organisms. Platypuses in particular can exavate extensive burrows on the banks of freshwater lakes and streams. Prototherians are hosts of various parasites (e.g. trypanosomes in platypuses and hepatozoans in echidnas).

- Ecosystem Impact: creates habitat.

Commensal/Parasitic Species

- Haemogregarinidae
- Trypanosoma binneyi

Economic Importance for Humans: Positive

One of the three species within Prototheria (Zaglossus bruinji) is eaten by the indigenous people of New Guinea. Hunting pressure has been so great that this species is now threatened with extinction.

Positive Impacts: food.

Economic Importance for Humans: Negative

Prototherians have no known negative impact on people, except perhaps for the pain a platypus can cause if it delivers venom with its spur. If unmolested, platypuses will not attack humans.

- Negative Impacts: injures humans (bites or stings).

Conservation Status

At least two of the three species of prototherians are threatened with extinction or were at some point in the recent past. Long-nosed echidnas are currently listed as endangered by the International Union for the Conservation of Nature (IUCN). Habitat destruction and heavy hunting pressure have severely limited populations of this species. In the past, platypuses were declining rapidly due to their overexploitation in the fur trade, but recent conservation efforts have helped populations rebound considerably.

Theria

Metatheria

Metatherian mammals, also known as marsupials, comprise around 272 species. They are an ancient group, very diverse in body form, and they occupy an enormous range of ecological niches. Today, most marsupials are found in Central and South America (around 70 species) and Australasia (around 200 species). Radiations took place on both of these continents during the Cenozoic, at a time when there were few placental competitors. Present marsupial faunas are very diverse, with some startling parallels with placental mammals (e.g., marsupials with similar morphologies and life histories as moles, anteaters, shrews, primates, carnivores, and many others). Some marsupial life histories and morphologies are seemingly without placental mammal parallels, for example, kangaroos. Past marsupial faunas were even more incredible. In Australia, for example, were rhinoceros-sized marsupial herbivores, kangaroos nearly 10 feet tall, and carnivorous lion-like forms with shearing teeth and retractile claws. In South America, where parallel radiations of large placental herbivores may have denied these herbivorous niches to marsupials, marsupials filled many carnivore niches (including a sabretooth marsupial "cat") and many rodent-like forms. It seems clear on both continents that invasion by placental mammals is correlated with a decline in number and diversity of marsupials. However, it is unclear whether placental mammals caused the disappearance of marsupials through competition or the apparent pattern of replacement is the result of random historical events.

Geographic Range

Today, most marsupials are found in Central and South America (around 70 species) and Australasia (around 200 species). One species is found in temperate North America (Didelphis virginianus). Metatheres diverged from the lineage leading to eutherian (placental) mammals by the Middle Cretaceous. Early diversification of metatheres is thought to have taken place in North America although, by the middle Miocene, the lineage became extinct on that continent, only reappearing around the time that North and South America regained contact in the Pliocene. The earliest marsupials are believed to have resembled North American opossums and other members of the family Didelphidae. A few fossil marsupials are known from Europe, Africa, and Asia, but this group was never well established on those continents.

Biogeographic Regions: Nearctic (native); neotropical (native); australian (native).

Habitat

From deserts and dry scrubland in Australia to tropical rain forests in South America, there are at least a few, and often many, species of Metatheria present. These animals occupy an enormous variety of terrestrial habitats throughout these two continents. The single species found in temperate North America (Didelphis virginia) naturally inhabits moist woodlands, but is common in towns and small cities. Metatheres have evolved to fill many niches in many habitats. Many species are fully terrestrial, many are arboreal, and at least one species, yapoks, is semi-aquatic.

- Habitat Region: stemperate; tropical; terrestrial; freshwater.
- Terrestrial Biomes:desert or dune; savanna or grassland; chaparral; forest; rainforest; scrub forest; mountains.
- Aquatic Biomes: lakes and ponds; rivers and streams.
- Wetlands: swamp.
- Other Habitat Features:urban; suburban; agricultural; riparian.

Physical Description

Marsupials differ from placental mammals in a number of important and obvious ways. The palate of marsupials is usually “fenestrated,” that is, it contains large gaps or spaces in its bony surface. The angular process of the dentary is inflected (bent) medially in almost all marsupials. The braincase is small and narrow. It houses a relatively small and simple brain compared to that of similar-sized placental mammals. The jugal is large, extending posteriorally so that it contacts, and forms part of, the glenoid fossa. The lacrimal canal is slightly anterior to the orbit so that it opens on the surface of the face rather than inside the orbital space. The bullae are sometimes not ossified. When they are, they are formed largely by extensions from the alisphenoid.

Tooth form varies considerably among species of marsupials, but an easy and reliable character for recognizing members of the group is that the number of incisors in the upper jaw is different from the number in the lower (except in one family, the Vombatidae). The number is equal in most (but not all) placental mammals. Also, the maximum number of incisors is 5/4, in contrast to 3/3 in placentals. The number of premolars and molars also differs between the groups (3/3 4/4 in marsupials, 4/4 3/3 in placental mammals), and the pattern of tooth replacement (milk teeth by adult teeth) differs, but these traits are difficult to use to recognize specimens. Postcranial skeletons of marsupials differ from those of placental mammals in that modern marsupials have epipubic bones in the body wall, projecting anteriorally from the pelvis. Epipubic bones are vestigial in recently extinct thylacines and were absent in at least one extinct group. The presence of epipubic bones is shared with monotremes.

Distinguishing among most of the orders and families of modern marsupials is not difficult. Two frequently-used characteristics are the conformation of the feet, and the number and position of the lower incisors. The second and third toes of syndactylous species are mostly enclosed in a sheath of skin and appear fused, except for the claws. Members of the orders Peramelemorphia and Diprotodontia are syndactylous. Others have separate toes, sometimes referred to as polydactylous. Members of the Diprotodontia and Paucituberculata have a pair of enlarged, forward-projecting (procumbent) lower incisors, a condition called diprotodonty. Other groups are polyprotodont, with numerous small and unspecialized lower incisors. Marsupial moles (Notoryctemorphia) are an unusual group, probably because of their extreme specialization to a fossorial mode of life. Their incisor morphology is not clearly diprotodont or polyprotodont and their feet are neither polydactylous nor syndactylous.

- Other Physical Features: endothermic; homoiothermic;bilateral symmetry.
- Sexual Dimorphism: sexes alike; male larger; sexes colored or patterned differently.

Reproduction

Mating systems of metatheres vary considerably. Many species are solitary throughout the year, only coming together to mate. This pattern of social behavior probably reflects promiscuous mating systems. Males of some species defend their access to several females. Koalas are an example of this polygynous mating behavior. Polygyny can also take the form of male dominance hierarchies in highly social species such as whiptail wallabies, which can live in groups of up to 50 individuals. Monogamy is also present within Metatheria. For example, *Petauroides volans* live in small family groups that consist of a mated pair and their offspring.

- Mating System: monogamous; polygynous; polygynandrous (promiscuous).

Marsupials and placental mammals differ strongly in their reproductive anatomy and pattern. In females, the reproductive tracts of marsupials are fully doubled. The right and left vaginae do not fuse to form a single body, as they do in all placental mammals, and birth takes place through a new median canal, the pseudovaginal canal. Right and left uteri also are unfused (varying degrees of fusion are found in placental mammals). Also, in the developing marsupial embryo, the arrangement of ducts that become the female reproductive tract is different in marsupials compared to placentals. In some (but by no means all) species of marsupials, females develop a pouch or marsupium in which the young are nursed. In males, the penis, like the female vagina, is bifid, or doubled. The scrotum lies in front of the penis instead of posterior to it, as in placental mammals.

Perhaps the most conspicuous difference between marsupials and placental mammals is in the degree of development of the young at birth. Marsupial young are tiny at birth; litters always weigh less than 1% of the mother's body weight and individual young sometimes weigh only a few milligrams. They are born after a very short gestation period (8 to 43 days, depending on species; always less or equal to the length of an estrus cycle), and in what seems to our placental-biased point of view to be an extraordinarily underdeveloped state. A placenta is formed in only a few species and, even in those, the gestation period is extraordinarily short. At the time newborn marsupials emerge from their mother's reproductive tract and crawl to the pouch, they are tiny and have just begun forming functional organs. The forelimbs are fairly well developed, as they are required for the young pull to themselves along the mother's belly by grasping hairs with the forelimbs, but the hindlimbs are mere paddles. The heart, kidneys, and lungs are all barely functional. Even the brain is at an early ontogenetic stage. Most development takes place in the pouch and the lactation period is prolonged.

It has been suggested that the marsupial pattern of reproduction is primitive for metatherian and eutherian mammals. Lillegraven argues that marsupial young must be born quickly, before the mother's immune system can respond the presence of foreign tissue in the form of a developing embryo. Most development takes place in the pouch, safe from maternal immune attack. Eutherians "solved" the problem of immune rejection through the evolution of a complex set of interactions that take place in the trophoblast, a part of the developing egg of eutherians that is not found in marsupials. Whether this is likely to be true and whether retaining a primitive style of reproduction suggests any kind of competitive inferiority has been hotly debated.

- Key Reproductive Features: iteroparous; gonochoric/gonochoristic/dioecious (sexes separate); sexual; viviparous.

Much of development in metatheres occurs after parturition while the young are nursing. In about 50 percent of marsupial species, young develop within the confines of a marsupium, or pouch. Weaning may take place after a year or more in some species (e.g., Macropus). Thus, female metatheres invest very little energy and resources into gestation, but lactation requires a substantial investment. The pouch itself (or protective folds of skin in many species) may be permanent, or may only develop at the onset of reproduction. In either case, resources must be devoted to producing and maintaining the structures that will protect the developing young.

Young generally do not associate directly with their mothers for much more than several weeks once they are fully independent of the pouch. This is generally true for both non-social and social species. In at least one species (Marcopus rufogriseus), extended associations between females and their independent young are known to reduce the success of future reproduction.

- Parental Investment: altricial; pre-hatching/birth; provisioning; female; pre-weaning/fledging; provisioning;female; protecting; female; pre-independence; protecting; female; post-independence association with parents.

Lifespan/Longevity

While some species of metatheres live only 1 to a few years, some species, such as coarse-haired wombats, have lived up to 26 years in captivity.

Behavior

Metatheria has undergone two major adaptive radiations (in Australia and South America) and exhibit a vast array of behavior as a consequence of evolving to fill a variety of ecological niches on these two continents. Thus, generalization of their behavior is difficult. Marsupials may be arboreal, terriculous, or fossorial and at least one species is semi-aquatic (yapoks or water opossums). Their locomotion may include walking, gliding, hopping, running and swimming. Marsupials have not evolved powered flight. Activity patterns also run the full spectrum of possibilities: marsupials may be diurnal, nocturnal or crespuscular. Seasonal behavior may include hibernation in some species, while others remain active throughout the year. Some species are social, while others are solitary.

- Key Behaviors: arboreal; Scansorial; Cursorial; Terricolous; fossorial; Glides; saltatorial; natatorial; diurnal; nocturnal; crepuscular; motile; hibernation; solitary; territorial; social.

Communication and Perception

As with mammals in general, vision, olfaction, hearing and touch are all important to varying degrees among species of Metatheria. Communication can take many forms as well. Some marsupials communicate with acoustic signals, particularly during mating or territorial encounters. Many species have conspicuous color patterns that may convey information about sex or species identity. Pheromones may also used in communication of reproductive receptivity.

- Communication Channels: visual; tactile; acoustic; chemical.
- Other Communication Modes: pheromones.
- Perception Channels: visual; tactile; acoustic; chemical.

Food Habits

Ominivory, insectivory, carnivory, and herbivory are all common food habits among Metatheria. Some groups that exhibit omnivory are American oppossums (Didelphidae), Australian possums (Phalangeridae) and bandicoots (Peramelidae). Insectivory and/or carnivory can be observed in several groups as well (e.g., Dasyuridae). Many marsupials, such as koalas (Phascolarctidae), wombats (Vombatidae), and kangaroos (Macropodidae), are strictly herbivorous.

- Primary Diet: carnivore; eats terrestrial vertebrates; insectivore; eats non-insect arthropods; herbivore; folivore; frugivore; granivore; Lignivore; nectarivore; omnivore.

Predation

Except for a very few species (e.g., the extinct thylacine wolf), marsupials are rarely top carnivores and thus are subject to predation by a host of mammalian, reptilian and avian predators wherever they occur.

Ecosystem Roles

With their great diversity of food habits, behavior and habitat use, metatheres can substantially impact their communities and ecosystems in a variety of ways. For example, metatheres may help pollinate plants, distribute seeds, or control pest populations. Most species are prey for other species and thus are an important component of many food webs. Species that dig burrows (e.g. wombats and marsupial moles) create habitat for other organisms and/or help aerate soil. Parasites of marsupials are certainly as diverse as their hosts.

- Ecosystem Impact: disperses seeds; pollinates; creates habitat; soil aeration.

Economic Importance for Humans: Positive

Humans benefit from metatheres in a variety of ways. Many are eaten as food or their body parts are used as some sort of resource (e.g., leather is made from kangaroo hides and koalas and brushtail possums were once taken for their fur). Some species that eat mice or insects can help control agricultural pests. Metatheres are valuable for the ecotourism industry--drawing many tourists to Australia. Some species are even kept as pets (e.g. *Petaurus* outside of Australia).

- Positive Impacts: pet trade; food; body parts are source of valuable material; ecotourism; controls pest population.

Economic Importance for Humans: Negative

Metatheres generally do not have large detrimental effects on human economics. Some herbivorous species may be minor crop pests. Wombats are extirpated in some areas because their burrows cause injury to livestock, or because European rabbits, which have become significant pests in Australia, use wombat burrows as dens.

Eutheria

In some taxonomic classification schemes, Eutheria is the term used for one of the three major branches or clades of mammals that includes extant members, the other two groups being the marsupials and the monotremes. Eutheria is often presented as an infraclass within the subclass Theria within the class Mammalia, with the other Infraclass within Theria being Metatheria (marsupials and their nearest ancestors). As used in this sense, Eutheria includes all extant placentals and those extinct mammals that share a more recent common ancestor with placentals than they do with Metatheria. The order Montremata (echidnas and platypuses) are often placed within the subclass Prototheria.

Extant eutherians, their last common ancestor, and all extinct descendants of that ancestor are placentals. Except for the Virginia opossum, which is a metatherian, all mammals indigenous to Europe, Africa, Asia, and North America north of Mexico are eutherians. There are over 1000 extant genera and over 4000 extant species of eutherians, ranging from shrews to whales and from aquatic cetaceans to the volant bats to the terrestrial primates.

However, not all classification schemes utilize the term Eutheria. For example, the popular scheme of McKenna and Bell recognize the two cohorts of Placentalia and Marsupialia, choosing not to use Eutheria. In this sense, Placentalia is used synonymous with Eutheria.

Eutherians are distinguished from noneutherians by various features of the feet, ankles, jaws, and teeth. One of the major differences between placental and nonplacental eutherians is that placentals lack epipubic bones, which are present in all other fossil and living mammals (marsupials and monotremes).

Mammals (class Mammalia) are a clade of endothermic amniotes distinguished from the reptiles and the birds by the possession of hair, three middle ear bones, mammary glands in females, and a neocortex (a region of the brain). The mammalian brain regulates body temperature and the circulatory system, including the four-chambered heart. The basic body type is a four-legged land-borne animal, but some mammals are adapted for life at sea, in the air, in the trees, or on two legs.

In some classifications, the mammals are divided into two subclasses (not counting fossils): the Prototheria (order of Monotremata) and the Theria, the latter composed of the infraclasses Metatheria and Eutheria. The marsupials constitute the crown group of the Metatheria and therefore include all living metatherians as well as many extinct ones; the placentals likewise constitute the crown group of the Eutheria.

While the classification of mammals at the family level has been relatively stable, different treatments at higher levels—subclass, infraclass, and order—appear in contemporaneous literature, especially for the marsupials. Much recent change has reflected the results of cladistic analysis and molecular genetics. Results from molecular genetics, for example, have led to the adoption of new groups such as the Afrotheria and the abandonment of traditional groups such as the Insectivora.

As with other mammals, eutherians are endotherms and amniote vertebrates, and have at most two generations of teeth, unlike reptiles in which there can be many generations of teeth. In mammals, the articular and quadrate, which composes the jaw joint in reptiles, is incorporated into the middle ear as the outermost two ear ossicles (malleus and incus; reptiles have only one ear ossicle, thte columnella). While in birds and reptiles, the large aorta leaving the heart bends to the right, in mammals it bends to the left.

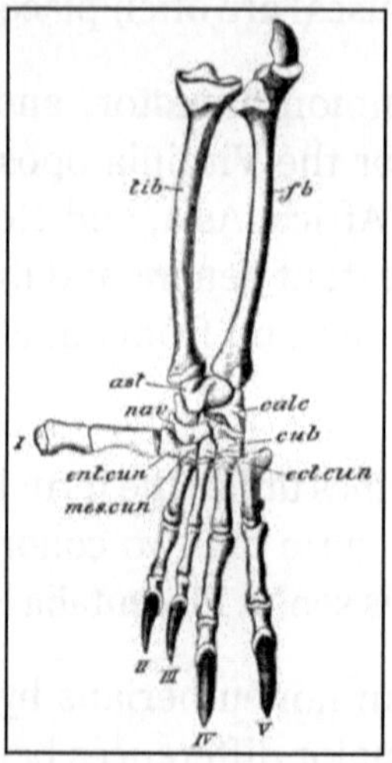

Bones of leg and foot of a mammal, Phalanger

The features of Eutheria that distinguish them from metatherians, a group that includes modern marsupials, are:

- An enlarged malleolus ("little hammer") at the bottom of the tibia, the larger of the two shin bones.
- The joint between the first metatarsal bone and the entocuneiform bone in the foot is offset further back than the joint between the second metatarsal and middle cuneiform bones—in metatherians these joints are level with each other.
- Various features of jaws and teeth.

Basic Design

Eutherians share with all other mammals some key innovations that differentiate them from other amniote vertebrates – Reptilia (including Aves). While in reptiles there can be many generations of teeth, in mammals there are at most two. Eutherians, if they have teeth, retain the ancestral mammal condition of two generations (deciduous and permanent) of teeth. Reptiles have a jaw joint composed of the articular (lower jaw) and quadrate (upper jaw), and have only one ear ossicle, the columnella. In all mammals the articular and quadrate become incorporated into the middle ear as the outermost two ear ossicles, the malleus and incus, respectively, which articulate with the intermost stapes (columnella). While prototherians lack teeth as adults, metatherians retain at most five upper and four lower incisors, one upper and one lower canine, three upper and three lower premolars, four upper and lower molars each. This condition is still found in the opossum, common to many areas of North America. Primitively, eutherians had a similar number of incisors and canines, but had five upper and lower premolars each and three upper and lower molars each. Except for placentals that have supernumerary teeth (e.g. some whales, armadillos, etc.), in extant placentals the number of teeth is at most three upper and lower incisors, one upper and lower canine, four upper and lower premolars, and three upper and lower molars. Except for one fewer upper molar, a domestic dog retains this pattern. Compared to reptiles, mammals have fewer skull bones through fusion and loss, although bones are variously emphasized in each of the three major mammalian taxa.

Physiologically, mammals are all endotherms of varying degrees of efficiency. They are also homeothermic with a relatively high resting temperature. These characteristics are also found in birds, but because of anatomical differences, the attainment of endothermy evolved convergently in mammals and birds. In mammals the large aorta leaving the heart bends to the left while in birds and their reptilian relatives the aorta bends to the right. Although both birds and mammals have diaphragms, they are formed very differently, again indicating convergent evolution.

Reproductively, mammals show all three major kinds of reproduction found in amniote vertebrates – oviparity or egg-laying, ovoviviparity where the embryo is retained internally by the mother but there is little maternal support, and euviviparity where the embryo is retained internally by the mother and much support is given by the mother. It is this last condition, euviviparity, that characterizes placentals. The name placental derives from the dominant extraembryonic structure of the same name found in this group. Both marsupials and placentals have a placenta but of considerably different structure. In marsupials two extraembryonic structures, the yolk sac and the chorion, fuse through part of their extent to form the choriovitelline placenta. In placentals, the allantois and chorion fuse to form the chorioallantoic placenta. Although the choriovitelline placenta of the marsupial compared to the chorioallantoic placenta of the placental does not produce as many hormones to sustain itself or provide as long a period of sustenance for the developing embryo, it should not be thought of as more primitive.

Rather, because the two kinds of placenta are formed differently they almost certainly evolved convergently.

Afrosoricida

Afrosoricida is an order of small African mammals that contains two extant families: the golden moles comprising the Chrysochloridae family and the tenrecs (and otter-shrews) comprising the Tenrecidae family. Members of this order can be quite diverse, but, in addition to being united on the basis of genetic studies, do share the presence of a cloaca (common opening for anal and urogenital tracts) and generally (but not in all species) zalambdodonty (with the upper molar characterized by one major v-shaped ridge or cusp). Modern golden moles are found in southern Africa and the tenrecs in Madagascar and western central Africa.

The term Afrosoricida means "looking like African shrews," reflecting some species' physical similarity with the true shrews of order Soricomorpha, the treeshrews of order Scandentia, and the elephant shrews or "sengis" (which are also restricted to Africa) of the order Macroscelidea. However, tenrec species are widely diverse phonologically and can resemble hedgehogs, mice, opossums, and even otters. They occupy aquatic, arboreal, terrestrial, and fossorial environments.

Members of Afrosoricida play a key role in food chains, consuming invertebrates and helping to control insect pest populations while being consumed by snakes, owls, and mammalian carnivores.

Golden moles (family Chrysochloridae) and tenrecs (family Tenrecidae) used to be classified in an order called Insectivora with other insectivorous mammals, such as shrews (family Soricidae), solenodons (family Solenodontidae), moles (family Talpidae), hedgehogs and gymnures (family Erinaceidae). This Insectivora order is now abandoned, as not all insectivorous mammals are closely related. Molecular evidence indicated that Chrysochloridae (golden moles) and Tenrecidae (tenrecs and otter-shrews) should be placed together in a new order called Afrosoricida, as identified and named by Stanhope et al. Afrosoricida now is generally considered part of the newly recognized clade or superorder known as Afrotheria, which includes the sengis (order Macroscelidea), elephants (Proboscidea), hyraxes (order Hyracoidea), manatees and dugons (order Sirenia), and aardvarks (order Tubulidentata). However, there is no strong morphological evidence to link the Afrosoricida together with other Afrotherians. The Afrosoricida are sometimes considered part of the Afroinsectiphilia, a clade within Afrotheria.

Some biologists use Tenrecomorpha as the name for the tenrec-golden mole clade, but Bronner and Jenkins argue that Afrosoricida is more appropriate, despite their misgivings about the similarity between the name "Afrosoricida" and the unrelated shrew subgenus *Afrosorex*.

Infraclass Eutheria: Placental Mammals

- Superorder Afrotheria:
    - Clade Afroinsectiphilia,
    - Order Afrosoricida,
    - Suborder Tenrecomorpha,
    - Family Tenrecidae: tenrecs and otter shrews; 30 species in 10 genera,
    - Suborder Chrysochloridea,
    - Family Chrysochloridae: golden moles; about 21 species in 9 genera,
    - Order Macroscelidea: elephant shrews,
    - Order Tubulidentata: Aardvark,
    - Clade Paenungulata,
    - Order Hyracoidea: hyraxes,
    - Order Proboscidea: elephants,
    - Order Sirenia: manatees and dugongs.

Golden Mole

Golden mole

Golden moles form the family Chrysochloridae. The golden moles bear a resemblance to both the true moles in family Talpidae and the marsupial moles in family Notoryctidae, including having small ears hidden by their fur, short tails, and eyes covered by skin. Their resemblance to the marsupial moles of Australia is so remarkable that, the marsupial/placental divide notwithstanding, arguments were once made that they were related, possibly because they are very primitive placentals and because of the many mole-like specializations. Golden moles are common in southern Africa. They live almost exclusively underground, beneath grassveld, forest, swamps, deserts, or mountainous terrain.

Golden moles range in size from about 8 to about 20 cm. Like several other burrowing mammals with similar habits, golden moles have short legs with powerful digging claws, very dense fur that repels dirt and moisture, and toughened skin, particularly on the head. Their eyes are non-functional and covered with skin and fur, the ears are just tiny openings, and, like the marsupial moles, they have an enlarged leather-like pad to protect their nostrils. Golden moles have muscular shoulders and an enlarged third claw on the forelimbs to aid digging, with no fifth digit and vestigial first and fourth digits; the hind feet retain all five toes and are webbed to allow efficient backward shoveling of the soil loosened with the front claws. Their malleus is very enlarged; their primary sense is that of touch, and they are particularly sensitive to vibrations that may indicate approaching danger. Both males and females have a single opening for the urogenital system (cloaca). They have a unique pair of bones in the occipital area of the skull called tabulars. The first incisor is enlarged.

While the desert species simply 'swim' through loose sand, all other species construct permanent burrows. The burrows are relatively complex in form, and may penetrate as far as a meter below ground. They include deep chambers for use as bolt-holes, and others as latrines. Excavated soil is pushed up to the surface as ridges or mole-hills, or is compacted into the tunnel walls. During extremely hot weather, Grant's golden mole will retreat to depths of around 50 cm and enter a state of torpor, thus conserving energy.

Because these mammals were previously thought to have originated in Gondwana, golden moles used to be regarded as rather "primitive" creatures: their low resting metabolic rate and their ability to switch off thermoregulation when inactive, however, are no longer regarded as indications that golden moles are undeveloped "reptilian mammals," but rather as essential adaptations to a harsh climate. By going into a torpor when resting or during cold weather, they conserve energy and reduce their need for food. Similarly, they have developed particularly efficient kidneys and most species do not need to drink water at all.

Females give birth to one to three hairless young in a grass-lined nest within the burrow system. Breeding occurs throughout the year. The adults are solitary, and their burrowing territory may be aggressively defended from intruders, especially where resources are relatively scarce.

Golden moles feed on small insects, which are located with the sense of hearing. Golden moles use a head-dipping behavior to detect the vibrations produced by grass mounts in order to position themselves within detection range of the head-banging alarms produced by their termite prey. Grant's golden mole (Eremitalpa granti) can cover 6 km each night looking for food.

Of the 21 species of golden mole, no fewer than 11 are threatened with extinction. The primary causes are sand mining, poor agricultural practices, increasing urbanisztion, and predation by domestic cats and dogs.

Tenrec

Greater hedgehog tenrec, Setifer setosus.

Tenrecs (some species of which are commonly known as "otter-shrews") comprise the family Tenrecidae. The are more numerous and diverse on Madagascar, but some are found in the western central areas of the African mainland. Tenrecs are widely diverse; as a result of convergent evolution, they resemble hedgehogs, shrews, opossums, mice, and even otters. They occupy aquatic, arboreal, terrestrial, and fossorial environments. Some of these species, including the greater hedgehog tenrec, can be found in the Madagascar dry deciduous forests. The three species in the subfamily Potamogalinae ("otter-shrews") are all aquatic forms.

As a rule, tenrecs tend to be small animals. The smallest species are the size of shrews, with a body length of around 4.5 cm (1.8 in), and weighing just 5 g (0.18 oz), while the largest, the common or tailless tenrec, is 25 to 39 cm (9.8 to 15 in) in length, and can weigh over 1 kilogram (2.2 lb).

Tenrecs do not have a pronounced body type, having evolved to fill the insect-eating niche in Madagascar. Their coat can vary from smooth to spiny and the coloration of the fur is generally dirt brown. Members of the subfamily Tenrecinae all have spines. As with golden moles, but unusual among placental mammals, the anus and urogenital tracts of tenrecs share a common opening, or cloaca, a feature more commonly seen in birds, reptiles, and amphibians. They have low body temperatures, sufficiently so that the males do not require a scrotum to cool their sperm as most other mammals do. (Golden moles also lack a scrotum.)

Most species are nocturnal and have poor eyesight. Their other senses are well developed, however, and they have especially sensitive whiskers. As with many of their other features, the dental formula of tenrecs varies greatly between species; they can have from 32 to 42 teeth in total. With the exception of the African otter shrews, the upper molars are zalambdodont; in the African otter shrews (subfamily Potamogalinae) the molars are dilambdodont. Unusual for mammals, the permanent dentition in tenrecs tends not to completely erupt until well after adult body size has been reached. This is one of several anatomical features shared by elephants, hyraxes, sengis, and golden moles (but apparently not aardvarks), consistent with their descent from a common ancestor.

species appear to be at least somewhat omnivorous, with invertebrates forming the largest part of their diets. The three species found on the African mainland (Potamogale velox, Micropotamogale lamottei, M. ruwenzorii) have more specialized diets, centered on their habitat in fast-running streams of the African tropics, from Liberia in the west to Lake Victoria in the east. One species from Madagascar, Limnogale mergulus, is also semiaquatic.

At least some species of tenrecs are social, living in multigenerational family groups with over a dozen individuals. Tenrecs have a gestation period of 50 to 64 days, and give birth to a number of relatively undeveloped young. While the otter shrews have just two young per litter, the tailless tenrec can have as many as 32, and females possess up to 29 teats, more than any other mammal.

Physical Characteristics

Apart from the obvious cloacal opening and the teeth structure, tenrecs tend to be small animals varying from 4cm to 39 cm in length. There is no pronounced body type since they have evolved to take over the insect-eating niche in Madagascar. However, based on the niche occupied, they look like shrews, hedgehogs or otters. There coat can vary from smooth to spiny and the coloration of the fur is generally dirt brown. Most species are also nocturnaland have poor eyesight. However, their whiskers are rather senstive and they can detect very minute vibrations in the ground to locate their prey.

Geographic Distribution

Tenrecs are an amazing example of speciation in isolated ecologial niches. Very much like the marsupials in Australia, many types of Tenrecs have colonized Madagascar and have radiated into niches occupied by rats, hedgehogs and even ant-eaters. These amazing creatures are found mostly in forest habitats but some species also inhabit arid regions in Madagascar. Their relatives, the golden moles live in the Namib desert, swimming in the sands, snacking on insects and generally having a merry old time. The only tenrecs left in mainland Africa are the otter shrews that live in the jungles of Congo and occupy the niche that an otter would.

Reproduction

Tenrecs have relatively long pregnancies (8 weeks) when compared to other insectivores like shrews(4 weeks). The size of a litter varies with each species depending on the type of climate and the lack or abundance of food. One species called the tailless tenrec can have the largest litter next to the naked mole rat. They can have upto 32 young.

Feeding Habits

This order is largely fond of insects and other invertebrates. Some of the larger species can dig into amphibians once in a while and aquatic species indulge in crayfish too. However, there is very limited specialization in eating habits otherwise.

Macroscelidea (Sengis)

Elephant shrews, or "sengis," are not shrews at all, despite their long inclusion by taxonomists within the order Insectivora, which does include shrews (family Soricidae).

The same and related genetic and morphological comparison studies support the inclusion of several African mammal orders into the superorder Afrotheria: the sengis (order Macroscelidea); elephants (Proboscidea); manatees and dugongs (order Sirenia); hyraxes (order Hyracoidea); aardvarks (order Tubulidentata); and Afrosoricida (or Tenrecomorpha), including golden-moles (family Chrysochloridae) and tenrecs and otter-shrews (family Tenrecidae).

Although still a widespread, vigorous family of mammals, the glory days of the Macroscelididae were in the past, when there were many more species, including an additional four families. The 15 species that remain are but leftovers of an extensive pan-African radiation of sengis that began as far back as the Eocene Epoch, then peaked during the Miocene and Pliocene Epochs (24 to 2 million years ago [mya]). By 2 mya, all but the present living sengi species had become extinct.

Studies of the teeth of fossil and modern sengis indicate that the earliest ancestral sengis were primarily or exclusively consumers of plant material, some forms gradually changing over the ages to a more insectivorous diet. Present-day sengis eat only animal food, mostly invertebrates, or a combination of that and plant food.

The earliest known fossils in the sengi line are Chambius kasserinensis from the early Eocene of Tunisia and Herodotius pattersoni from the Late Eocene of Egypt (Eocene Epoch: 55 to 34 mya). The dental anatomy of these and other sengi fossils support (but do not confirm) a common ancestry of sengis with condylarths, primitive, extinct ungulate animals that gave rise to numerous lines of more recent and modern ungulates.

There are two subfamilies within the single surviving family of Macroscelididae: Rhynchocyoninae, the giant sengis, with the single genus Rhynchocyon; and Macroscelidinae, the soft-furred sengis, with the genus Elephantulus and the monotypic genera Petrodromus and Macroscelides.

Physical Characteristics

Sengis can charm the viewer with their rather humorous and endearing appearance and behavior. The sight of a sengi calls to mind a mouse with long, spindly legs and a mobile, slightly downturned snout varying in length among species. The long legs enable a sengi to walk, trot, run, or hop like a large, long-legged mammal, rather than with the scampering motions common among other small mammals. The body is compact, the head large in proportion. The coat is soft and full. The eyes are large, dark, and limpid, and the long proboscis moves continually in hesitant, circular twitchings. The effect is cute and slightly cartoonish.

The different sengi species share similar body proportions, ranging from mouse-sized to rabbit-sized. The Rhynchocyon species are the largest, with an adult head and body length of 9–12.5 in (235–315 mm), the tail adding another 7.5–10 in (190–263 mm). Adult weights can reach 14.5–15.5 oz (408–550 g). An example of a smaller sengi species is the four-toed sengi (Petrodromus tetradactylus), with an adult head and body length of 7–8.5 in (185–220 mm), tail length of 5–7 in (130–180 mm), and weight of 5–9 oz (150–280 g), but still large in comparison to the round-eared sengi (Macroscelides proboscideus) and other Elephantulus species, that are smaller and more mouselike in appearance, with adult head-and-body lengths of 3.5–6 in (90–145 mm), tail lengths of 3–6.5 in (80–165 mm), and weights of 1–2.5 oz (25–70 g).

Larger sengi species tend toward bright colors and patterns, while smaller species display more camouflaging browns and grays, often closely matching the soil color of a particular area. There is little sexual dimorphism throughout the family.

Sengi limbs feature long bones, the hindlimbs longer than the forelimbs, for cursorial (running) and ricochetal (hopping weith the hind legs) locomotion. The tibia and fibula (lower bones of the hindlimbs) are long and fused, the metatarsals are lengthened, and the ulna and radius (lower bones of the forelimbs) are also long, thus lengthening the stride and contributing to high-speed running and jumping.

Numbers and arrangements of digits vary throughout the family, from the traditional 10 fingers-10 toes arrangement to various reductions and modifications. The long tail is furless or sprouts bristles that vary in density and texture among species.

A sengi's mobile, somewhat flexible snout, which inspired the "elephant" word in the old common name, is both sense organ and tool. When not foraging, a sengi continually moves its snout in a subdued, circular motion from the base, seeking scents. When foraging, a sengi pokes its snout into crevices or leaf litter, sniffing for food. The nostrils are located at the forward end of the snout. Long sensory vibrissae, or whiskers, arise from the base of the snout.

The senses of smell, vision and hearing are well-developed and highly tuned. The eyes are conspicuously large, dark, and limpid. Most species have a pale ring around each eye. The ears are large in proportion to the head, and in the Elephantulus species, the auditory bulla is almost grotesquely enlarged, to enhance the animal's already acute hearing abilities. The braincase is relatively large, and more complex than that of similarly-sized insectivores.

Sengi bodies are riddled with scent glands, at the base of the tail, soles of the feet, chest, behind the ears, at the corners of the mouth and in the genital and anal regions, with which they mark territory.

Metabolic rates among sengi species are like those in similarly small mammals, not lower or fluctuating, as in shrews. A few sengi species, such as the North African sengi (Elephantulus rozeti), can adjust metabolism and activity to changes in their environments, going into torpor when temperatures or food availability go low.

A sengi lives from one to five years in the wild. The record longevity, held by a captive Bushveld sengi (E. intufi), is eight years and nine months.

Distribution

All sengi species are confined to the African continent, including, for a few species, the island of Zanzibar (but not including Madagascar). The three Rhynchocyon species live in central and eastern Africa. The four-toed sengi (Petrodromus tetradactylus) is widespread through eastern and southern Africa. The round-eared sengi (Macroscelides proboscideus) is found only in southern Africa. A majority of the ten Elephantulus species occur in southern Africa, followed by eastern Africa, excepting the North African sengi (Elephantulus rozeti), found only in Morocco, Algeria, Tunisia, and Libya, separated by the Sahara Desert from all the other sengi species.

Habitat

Genus Rhynchocyon is mostly confined to tropical lowland and montane forests, while all other sengi species live in dry savannas, scrub forest and brush, rocky outcrops, and deserts.

Behavior

Sengis are alert, high-strung creatures with hair-trigger senses and reactions, always primed for escape, fleeing into hiding at the least threat. The enemies of sengis are legion. Snakes, birds of prey, and carnivorous mammals are known predators of sengis. Accordingly, the sengi lifestyle balances the vulnerability of small size, short life-span and low reproductive rate with keen senses, lightning-swift reflexes, and alterations that the sengi makes in its territory to ensure maximum safety.

Sengis sleep or hide in tucked-away shelters, such as rock crevices, burrows, nests and depressions of their own making, or burrows abandoned by other small mammals. Sengimade burrows often include an inconspicuous emergency exit.

Individuals may live alone (except during mating), or as monogamous, pair-bonded male and female, or in small groups. Whether a loner, or one of a pair or group, a sengi patrols its territory constantly. A male-female bonded pair defends its territory "sex-specifically" against other members of its species, i.e., males confront and chase off intruding males, females do likewise to trespassing females.

Active times of day and night vary among species. The Rhynchocyon species are diurnal, while other species tend toward crepuscular (twilight) or nocturnal activity. Some diurnal species switch to nocturnal foraging in exceptionally hot weather, on adequately moonlit nights, or if overly harassed by daytime predators.

Many of the soft-furred sengi species make and maintain elaborate trail systems through leaf litter or grass, with strategically located hiding spots along the way for quick cover from threats. Males and females scent-mark stones and twigs along the trail systems and deposit identifying little heaps of excrement where their path crosses a path in an abutting territory. The animals fastidiously maintain the trails, booting off intruding pebbles, twigs, and leaves with their forelimbs. This behavior keeps the trails open and ready for the sengi to flee along toward shelter in response to threats.

A trail-making sengi marks its trails with little heaps of excrement wherever its path crosses the path of an abutting, same-species territory. If individuals of abutting territories meet at the crossroads, they interract with elaborate threat rituals, ending in a stalemate and truce, or a fight.

Unlike the soft-furred sengi species, the giant sengi species seldom maintain trail systems, but an individual or pair will build a network of leaf nests throughout their territory. The giant species scent-mark their territories but never resort to excremental signatures. There are no elaborate rituals if individuals of abutting territories meet. One simply and pointedly chases off the other, finishing with a nip by incisors to the fleeing hindquarters.

Sengi limbs are built for walking, trotting, and high-velocity running and hopping. Most species go to hopping mode only occasionally, while all are quadrupedal walkers and runners. Sengis walk and run in digitigrade fashion, i.e., on the tips of their fore-digits and hind-digits. When threatened, they prefer to run for a hiding spot, but will go into hopping mode with their hind limbs, tail extended, in extreme danger.

In open country, soft-furred sengis often sunbathe just out-side their home shelters, sitting on their haunches, their senses always alert for the least inkling of a threat, at which point they instantly rouse and flee into hiding.

Some sengi species also "sand-bathe," wallowing in dry sand, a behavior noted in other rodent species. Sand-bathing both scent-marks the immediate territory and serves to clean the fur of accumulated oil, dandruff, and dirt. Established sand-bathing spots are scattered throughout a sengi trail system. Soft-furred sengis also clean their fur with their tongues and scratch with their hind legs, able thereby to reach all parts of their bodies.

Sengis are not particularly vocal creatures. Elephantulus and Rhynchocyon species make squeaking sounds, and drum the leaf litter with their hind feet. The four-toed sengi (Petrodromus tetradactylus) makes a cricketlike call. The golden-rumped sengi (Rhynchocyon chrysopygus), the four-toed sengi, and Elephantulus species all rap the ground with their hind feet to sound alarms or for other reasons, doing so in species-specific patterns that vary in regular or irregular rhythms, intervals between each rap, number of raps in a set, and number of sets in a series. Rhynchocyon species and Petrodromus tetradactylus make similar warning noises by rapping their tails against the ground.

Feeding Ecology and Diet

Digestion in sengis is similar to that of small insectivorous mammals, although ancestral sengis were herbivorous. All sengis prey on invertebrates, mostly insects, while most supplement this fare with fruits, seeds, and green plant material, the one exception being Rhynchocyon chrysopygus, the golden-rumped sengi, which eats only animals, mostly invertebrates. Smaller species of sengis feed mainly on ants and termites.

The sengi digestive tract includes a caecum (analogous to the appendix in humans). This organ has been little studied in the Macroscelidae, and may not be functional in all species. In species in which it does function, the caecum is full of bacteria that break down cellulose in ingested plant matter. The functional caecum is another legacy from the purely herbivorous diets of sengi ancestors. The caecum of some sengi species may also be used to store water for hot, dry months.

Reproductive Biology

Mating behavior varies according to whether the male and female of a sengi species are monogamous (for life) or solitary. Males of solitary species, or young males of monogamous species on their first mating run, go in search of rutting females of their particular species by sensing olfactory cues left in scent markings left by females in strategic spots. When a male and female of a solitary species have found each other, they stay together for several days, mate, then go their separate ways. Pair-bonded couples remain together as long as both are alive. They may be strictly monogamous or they may mate with other individuals while always reassuming the original pair relationship. Males take little or no part in direct care of the young.

A female sengi carries four or six mammae, depending on species. The reproductive systems of female sengis, during mating, show polyovulation, in which anywhere from a dozen to 100 egg cells are released during ovulation, most of which become.

Fertilized by male sperm and begin cell division, but only one, two, or three fertilized eggs will implant in the uterus, while the rest are expelled. This may be a holdover from past ages when sengi ancestors bore larger litters, as do most mammal species in their size range.

After a gestation period of about 50 days, sengis are born in small litters of one, two, or rarely, three or four young, which enter the world precocial, i.e., with their eyes open, full coats of fur, and able to move about and explore within a few days, often a few hours, of birth. The mother leaves the infants alone most of the time, in a shelter separate from the parents' shelter, coming by only to nurse at fixed intervals, a behavior known as absentee parental care. After five days the mother starts feeding the young mashed insects that she stuffs in her cheek pouches, in addition to her milk.

In about two weeks, the young emerge from their shelter, effectively weaned and able to forage, although they will remain with the mother for three or four weeks, accompanying her as she forages. Eventually, within a month to two months after weaning, the young strike off on their own, or are driven from the territory by the parents, to establish territory of their own, reaching sexual maturity by 40–50 days. A female sengi may produce several litters per year.

Primate

Primate is any mammal of the group that includes the lemurs, lorises, tarsiers, monkeys, apes, and humans. The order Primates, with its 300 or more species, is the third most diverse order of mammals, after rodents (Rodentia) and bats (Chiroptera). Although there are some notable variations between some primate groups, they share several anatomic and functional characteristics reflective of their common ancestry. When compared with body weight, the primate brain is larger than that of other terrestrial mammals, and it has a fissure unique to primates (the Calcarine sulcus) that separates the first and second visual areas on each side of the brain. Whereas all other mammals have claws or hooves on their digits, only primates have flat nails. Some primates do have claws, but even among these there is a flat nail on the big toe (hallux). In all primates except humans, the hallux diverges from the other toes and together with them forms a pincer capable of grasping objects such as branches. Not all primates have similarly dextrous hands; only the catarrhines (Old World monkeys, apes and humans) and a few of the lemurs and lorises have an opposable thumb. Primates are not alone in having grasping feet, but as these occur in many other arboreal mammals (e.g., squirrels and opossums), and as most present-day primates are arboreal, this characteristic suggests that they evolved from an ancestor that was arboreal. So too does primates' possession of specialized nerve endings (Meissner's corpuscles) in the hands and feet that increase tactile sensitivity.

Old World and New World monkeys.

As far as is known, no other placental mammal has them. Primates possess dermatoglyphics (the skin ridges responsible for fingerprints), but so do many other arboreal mammals. The eyes face forward in all primates so that the eyes' visual fields overlap. Again, this feature is not by any means restricted to primates, but it is a general feature seen among predators. It has been proposed, therefore, that the ancestor of the primates was a predator, perhaps insectivorous. The optic fibres in almost all mammals cross over (decussate) so that signals from one eye are interpreted on the opposite side of the brain, but, in some primate species, up to 40 percent of the nerve fibres do not cross over. Primate teeth are distinguishable from those of other mammals by the low, rounded form of the molar and premolar cusps, which contrast with the high, pointed cusps or elaborate ridges of other placental mammals. This distinction makes fossilized primate teeth easy to recognize. Fossils of the earliest primates date to the Early Eocene Epoch (56 million to 40 million years ago) or perhaps to the Late Paleocene Epoch (59 million to 56 million years ago). Though they began as an arboreal group, and many (especially the platyrrhines, or New World monkeys) have remained thoroughly arboreal, many have become at least partly terrestrial, and many have achieved high levels of intelligence.

Species of lemurs: Lemurs (suborder Strepsirrhini).

General Considerations

Size Range and Adaptive Diversity

Members of the order Primates show a remarkable range of size and adaptive diversity. The smallest primate is Madame Berthe's mouse lemur (Microcebus berthae) of Madagascar, which weighs some 35 grams (one ounce); the most massive is certainly the gorilla (Gorilla gorilla), whose weight may be more than 4,000 times as great, varying from 140 to 180 kg (about 300 to 400 pounds).

Western lowland gorilla (Gorilla gorilla gorilla), male.

Primates occupy two major vegetational zones: tropical forest and woodland–grassland vegetational complexes. Each of these zones has produced in its resident primates the appropriate adaptations, but there is perhaps more diversity of bodily form among forest-living species than among savanna inhabitants. One of the explanations of this difference is that it is the precise pattern of locomotion rather than the simple matter of habitat that governs overt bodily adaptations. Within the forest there are a number of ways of moving about. An animal can live on the forest floor or in the canopy, for instance, and within the canopy it can move in three particular ways: (a) by leaping—a function principally dictated by the hind limbs; (b) by arm swinging (brachiation)—a function particularly of the forelimbs; (c) by quadrupedalism—a function equally divided between the forelimbs and the hind limbs. On the savanna, or in the woodland-savanna biome, which substantially demands adaptations for ground-living locomotion rather than those for tree-living, the possibilities are limited. If bipedal humans are discounted, there is a single pattern of ground-living locomotion, which is called quadrupedalism. Within this category there are at least two variations on the theme: (a) knuckle-walking quadrupedalism, and (b) digitigrade quadrupedalism. The former gait is characteristic of the African apes (chimpanzee and gorilla), and the latter of baboons and macaques, which walk on the flats of their fingers. After human beings, Old World monkeys of the subfamily Cercopithecinae are the most successful colonizers of nonarboreal habitats.

West African, or masked, chimpanzee (Pan troglodytes verus).

Primates are a homogeneous group morphologically, and it is only in the realm of behaviour that differences between primate taxa are clearly discriminant. It can be said that the most successful primates (judged in terms of the usual criteria of population numbers and territorial spread) are those that have departed least from the ancestral pattern of structure but farthest from the ancestral pattern of behaviour. "Manners makyth man" is true in the widest sense of the word; in the same sense, manners delineate primate species.

Distribution and Abundance

The nonhuman primates have a wide distribution throughout the tropical latitudes of Africa, India, Southeast Asia, and South America. Within this tropical belt, which lies between latitudes 25° N and 30° S, they have a considerable vertical range. In Ethiopia the gelada (genus Theropithecus) is found living at elevations up to 5,000 metres (16,000 feet). Gorillas of the Virunga Mountains are known to travel across mountain passes at altitudes of more than 4,200 metres when traveling from one high valley to another. The howler monkeys of Venezuela (Alouatta seniculus) live at 2,500 metres in the Cordillera de Merida, and in northern Colombia the durukuli (genus Aotus) is found in the tropical montane forests of the Cordillera Central.

Adult male mountain gorilla (Gorilla gorilla beringei) in Virunga National Park, Democratic Republic of the Congo.

In habitat, primates are predominantly tropical, but few species of nonhuman primates extend their ranges well outside the tropics. The Barbary "ape" (Macaca sylvanus) lives in the temperate forests of the Atlas and other mountain ranges of Morocco and Algeria. Some populations of rhesus monkey (M. mulatta) extended until the middle of the 20th century to the latitude of Beijing in northern China, and the Tibetan macaque (M. thibetana) is found from the warm coastal ranges of Fujian (Fukien) province to the cold mountains of Sichuan (Szechwan). One of the most remarkable, however, is the Japanese macaque (M. fuscata), which in the north of Honshu lives in mountains that are snow-covered for eight months of the year; some populations have learned to make life more tolerable for themselves by spending most of the day in the hot springs that bubble out and form pools in volcanic areas. Finally, two western Chinese species of snub-nosed monkey, the golden (Rhinopithecus roxellana) and black (R. bieti), are confined to high altitudes (up to 3,000 metres in the case of the former and to 4,500 metres in the latter), where the temperature drops below 0 °C (32 °F) every night and often barely rises above it by day.

Japanese macaques groom each other at a spa in Nagano prefecture, east-central Honshu, Japan.

Although many primates are still plentiful in the wild, the numbers of many species are declining steeply. According to the International Union for Conservation of Nature (IUCN), more than 70 percent of primates in Asia and roughly 40 percent of primates in South America, in mainland Africa, and on the island of Madagascar are listed as endangered. A number of species, particularly the orangutan, the gorilla, some of the Madagascan lemurs, and some South American species, are in serious danger of extinction unless their habitats can be preserved in perpetuity and human predation kept under control. The populations of several species number only in the hundreds, and in 2000 a subspecies of African red colobus monkey (Procolobus badius) became the first primate since 1800 to be declared extinct.

Red-bellied lemur (Eulemur rubriventer) in the eastern Madagascar rainforest near Ranomafana.

Male red colobus monkey (Piliocolobus badius temminckii).

In the midst of these declines, the populations of some critically endangered primate species have increased. Concerted efforts to breed a type of marmoset, the golden lion marmoset (or golden lion tamarin; Leontopithecus rosalia), in captivity have been successful; reintroduction of that species into the wild continues in Brazil. The estimated number of western lowland gorillas, a species thought to be critically endangered, increased when a population of more than 100,000 was discovered in 2008 in the swamps of the Lac Télé Community Reserve in the Republic of the Congo.

Reproduction and life cycle

The stages of the life cycle of primates vary considerably in duration. Among the most primitive members of the group, these stages are broadly comparable to those of other mammals of similar size. Higher in the phylogenetic scale, they are substantially extended. The greatest difference is in the duration of the infant and juvenile stages combined; the least is in the gestation period, which, despite the general belief, cannot be consistently correlated with adult body size. Gibbons, which weigh considerably less than macaques, have a 20 percent longer gestation period.

Rhesus monkeys (Macaca mulatta).

The clear trend toward prolongation of the period of juvenile and adolescent life is probably to be associated with the corresponding trend toward a progressive elaboration of the brain. The extended period of adolescence means that the young remain under adult (primarily maternal) surveillance for a long period, during which time the juvenile acquires, by example from its mother and peers, the knowledge that will allow it to become properly integrated as a fully adult member of a complicated social system. One might therefore expect a close correlation between the period of adolescence, the brain size, and the complexity of the social system; and, insofar as the latter factor can be assessed, this appears to be the case.

Breeding Periods

The reproductive events in the primate calendar are copulation, gestation, birth, and lactation. Owing to the long duration of the gestation period, these phases occupy the female primate (among higher primates anyway) for a full year or more; then the cycle starts again. The female does not usually come into physiological receptivity until the infant of the previous pregnancy has been weaned.

Ring-tailed lemurs (Lemur catta) sitting in a tree, Madagascar.

Most lemurs and lorises show one or more discrete breeding seasons during the year, during which time they may undergo more than one reproductive estrous cycle (i.e., period of sexual activity). The breeding seasons are separated by periods of anestrus, which in bush babies and mouse lemurs are accompanied by changes in the skin of the external genitalia (vulva), which closes over, completely sealing the vagina. When living in the wild in the Sudan, the lesser bush baby (Galago senegalensis) has an estrus that occurs only twice yearly, during December and August. In captivity, however, breeding seasons may occur at any period in the year. In the wild, birth seasons are closely correlated with the prevailing climate, but in captivity under equable laboratory conditions, this consideration does not apply. For instance, in its native Madagascar, the ring-tailed lemur (Lemur catta) has only a single breeding season during the year, conception occurring in autumn (April) and births taking place in late winter (August and September). However, in zoos in the Northern Hemisphere, a seasonal inversion occurs in which the birth period shifts to late spring and early summer. These examples indicate the influence of environmental factors on the timing of the birth seasons.

Bush babies or galagos (Galago senegalensis).

Reproductive cycles in tarsiers, apes, and many monkeys continue uninterrupted throughout the year, though seasonality in births is characteristic mainly of monkey species living either outside the equatorial belt (5° north and south of the Equator) or at high elevations in equatorial regions, where dry seasons and seasonal food shortages occur. Seasonality of births in macaques (genus Macaca species) has been documented in Japan, on Cayo Santiago in the Caribbean (where an introduced population thrives under seminatural conditions), and in India. Observations of langurs in India and Sri Lanka, of geladas in Ethiopia, and of patas monkeys in Uganda have also demonstrated seasonality in areas with well-marked wet and dry seasons. Those within the equatorial belt tend to display birth peaks rather than birth seasons. A birth peak is a period of the year in which a high proportion of births, but not by any means all, are concentrated. Equatorial primates such as guenons, colobus monkeys, howlers, gibbons, chimpanzees, and gorillas might be expected to show a pattern of births uniformly distributed throughout the year, but population samples are as yet too small to make this assumption, and some equatorial monkeys, such as squirrel monkeys (genus Saimiri), are strictly seasonal breeders. Even in humans, there is evidence of high birth peaks. In Europe, the highest birth rates are reached in the first half of the year; in the United States, India, and countries in the Southern Hemisphere, in the second half. This may, however, be a cultural rather than an ecological phenomenon, for marriages in certain Western countries reach a peak in the closing weeks of the fiscal year, a fact that undoubtedly has some repercussions on the birth period.

Gestation Period and Parturition

The period during which the growing fetus is protected in the uterus is characterized by a considerable range of variation among primate species, but it shows a general trend toward prolongation as one ascends the evolutionary scale. Mouse lemurs, for example, have a gestation period of 54–68 days, lemurs 132–134 days, macaques 146–186 days, gibbons 210 days, chimpanzees 230 days, gorillas 255 days, and humans (on the average) 267 days. Even small primates such as bush babies have gestations considerably longer than those of nonprimate mammals of equivalent size, a reflection of the increased complexity and differentiation of primate structure compared with that of nonprimates. Although in primates there is a general trend toward evolutionary increase in body size, there is no absolute correlation between body size and the duration of the gestation period. Marmosets, for example, are considerably smaller than spider monkeys and howler monkeys but have a slightly longer pregnancy (howler monkeys 139 days, "true" marmosets 130–150 days).

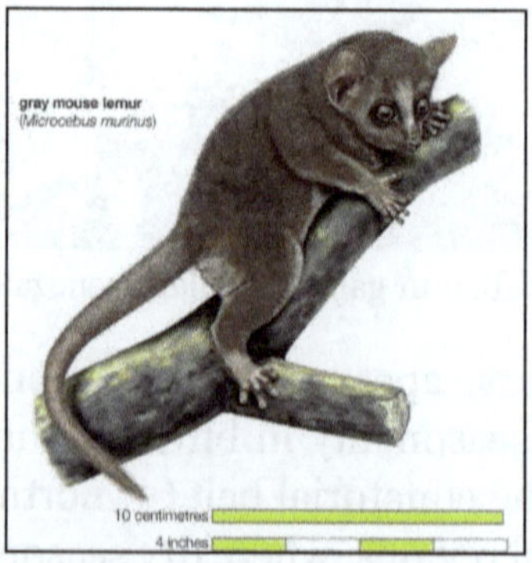

An extraordinary and somewhat inexplicable difference exists between the dimensions of the pelvic cavity and the dimensions of the head of the infant at birth in monkeys and humans on the one hand, and apes on the other. The head of the infant ape is considerably smaller than the pelvic cavity, so birth occurs easily and without prolonged labour. When the head of the infant monkey engages in the pelvis, the fit is exact, and labour may be a prolonged and difficult affair, as it is generally with humans. Human parturition, however, is generally a much more extended process than that of monkeys. Like the human infant, the monkey is born head first. Twin births are rare in most monkeys and apes, but marmosets and some lemurs and lorises habitually produce twins.

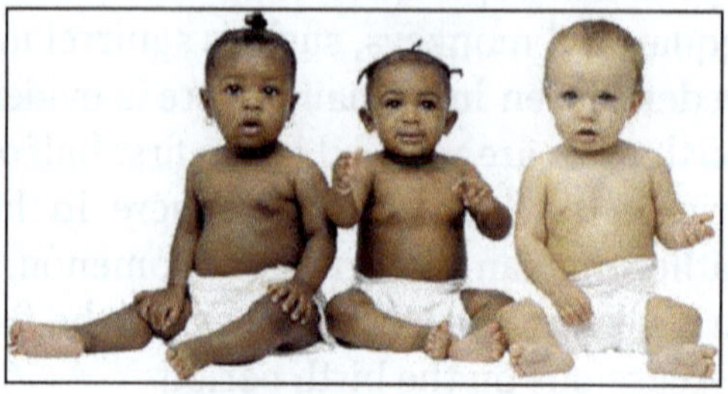
Three babies in diapers.

The degrees of maturation and mother dependency at birth are obviously closely related phenomena. Newborn primate infants are neither as helpless as kittens, puppies, or rats nor as developed as newborn gazelles, horses, and other savanna-living animals. With a few exceptions, primate young are born with their eyes open and are fully furred. Exceptions are mouse lemurs (Microcebus), gentle lemurs (Hapalemur), and ruffed lemurs (Varecia), which bear more helpless (altricial) infants and carry their young in their mouth. Primate life being peripatetic, it is axiomatic that the infants must be able to cling to the mother's fur; just a few species (again, mouse lemurs and ruffed lemurs and a few others) leave their infants in nests while foraging, and lorises "park" their young, leaving them hanging under branches in tangles of vegetation. The young of higher primates have grasping hands and feet at birth and are able to cling to the maternal fur without assistance; only humans, chimpanzees, and gorillas need to support their newborn infants, and humans do so longest.

Anubis, or olive, baboon (Papio anubis).

It seems likely that the difference between the African apes and humans in respect to postnatal grasping ability is related to the acquisition in man of bipedal walking. One of the anatomic correlates of the human gait is the loss of the grasping function of the big toe, which is aligned in parallel with the remaining digits. Such an arrangement precludes the use of the foot as a grasping extremity. The human infant—and to a lesser degree the gorilla infant—must depend largely on its grasping hands to support itself unaided. The fact that humans are habitually bipedal and that, consequently, the hands are freed from locomotor chores may also be a contributory factor; the human mother can move about and at the same time continue to support her infant. Selection for postnatal grasping, therefore, has not had the high survival value in humans that it has in nonhuman primates, in which the survival of the infant depends on its ability to hold on tightly. On the other hand, it is well known that newborn human infants can support their own weight, for short periods, by means of their grasping hands. Clearly then, adaptations for survival are not wholly lacking in the human species. Perhaps cultural factors have had the effect of suppressing natural selection for early infant grasping ability. The first factor may be the social evolution of a division of labour between the sexes and a fixed home base, which has allowed the mother to park her infant with other members of the family as babysitters. A second factor may be more peripatetic communities, in which the invention of infant-carrying devices, such as the papoose technique of North American Indians, has made it unnecessary for the infant to support itself. Whatever the biological or cultural reasons, the human infant is more helpless than the young of all other primates.

Once the primate infant has learned to support itself by standing on its own two (or four) feet, the physical phase of dependency is over; the next phase, psychological dependency, lasts much longer. The human child is metaphorically tied to its mother's apron strings for much longer periods than are the nonhuman primates.

Growth and Longevity

The prolongation of postnatal life among primates affects all life periods, including infantile, juvenile, adult, and senescent. Although humans are the longest-lived members of the order, the potential life span of the chimpanzee has been estimated at 60 years, and orangutans occasionally achieve this in captivity. The life span of a lemur, on the other hand, is about 15 years and a monkey's 25–30 years.

The characteristic growth spurts of human infants in weight and height also occur in nonhuman primates but start earlier in the postnatal period and are of shorter duration. Primates differ from most nonprimate mammals by virtue of a delayed puberty in both sexes until growth is nearly complete; in humans, the peak of the growth spurt in boys comes slightly after the sexual maturity, whereas in girls the growth spurt precedes menarche. There is some controversy over the very existence of an adolescent growth spurt in nonhuman primates. In some species, males are very much larger than females; this extra growth occurs long after sexual maturity and rather rapidly, so it is possibly equivalent to the human growth spurt. The most remarkable case of such post-mature growth is seen in orangutans. A male can mature physically in his early teens, or he can spend as much as 20 years as a subadult and then suddenly, within a year, almost double his weight and develop the cheek flanges characteristic of full maturity. It appears that this is related to social conditions; in proximity to a full-grown, dominant male, a subadult male's development will remain suppressed, and when the dominant male moves away (or, in a zoo, is removed from the vicinity), the subadult undergoes a flush of testosterone and matures rapidly.

Male orangutan (Pongo pygmaeus).

Locomotion

Primate locomotion, being an aspect of behaviour that arises out of anatomic structure, shows much of the conservativeness and opportunism that generally characterizes the order. Primates with remarkably few changes in their skeletons and musculature have adopted a bewildering variety of locomotor patterns. The "natural" habitat of primates—in the historical sense—is the canopy of the forest. Although many primates have adopted the ground as their principal foraging area during the day, given the opportunity they will return to the trees to sleep at night. Trees provide cover from the climate and protection from predators; they are of course also a source of food. Only the gelada, the hamadryas baboon of the mountainous regions of Ethiopia, and the chacma baboon, which lives on the rocky coast of the Cape of Good Hope, South Africa, are ground sleepers; yet even these animals seek the protection of the cliffs and rocky precipices of their habitats at night. No primate sleeps totally unprotected; as a consequence of their relative immunity from predation, primates are heavy sleepers.

Chinese white-cheeked gibbons (Hylobates leucogenys), male and female.

Four Types of Locomotion

The essential arboreality of primates has guaranteed the relative uniformity of the locomotor apparatus. Even humans, who have long since abandoned the trees as their principal lodging place, have only partially lost the physical adaptations for tree climbing; their hands, in particular, remain in the arboreal mold. Only the feet have lost their primitive prehensility in adapting to bipedal walking. Primate locomotion can be classified on behavioral grounds into four major types: vertical clinging and leaping, quadrupedalism, brachiation, and bipedalism. Within these major categories, there are a number of subtypes, and within these subtypes, there are an infinite number of variations between species and, by virtue of individual variability, within species. The differences between the four major categories lie principally in the degree to which the forelimbs and hind limbs are used to climb, swing, jump, and run.

Vertical clinging and leaping, for instance, is primarily a function of the hind limbs, as is bipedalism, whereas brachiation is performed exclusively with the forelimbs. Quadrupedalism involves both forelimbs and hind limbs, of course, although not to an equal extent. Some quadrupeds are hind limb-dominated; in others, the forelimb and the hind limb are equally important. The hind limb-dominated primates, such as the langurs and colobus monkeys, employ a large element of leaping in their movements, a less-notable feature of the more generalized quadrupeds such as guenons. The quadrupedal category is inevitably somewhat of a grab bag, and the gaits included in it have not yet been studied critically. One subtype, here designated as slow climbing, differs profoundly from the other subtypes of the category, being somewhat ponderous and devoid of elements of leaping or jumping. The species in this category are lorises and pottos, all of which are arboreal and nocturnal.

As many authorities who have studied locomotion in free-ranging primate species have pointed out, the classifications of locomotion into categories is a somewhat artificial procedure. A chimpanzee shows a variety of different gaits according to the circumstances of the environment: quadrupedalism (knuckle walking), climbing, bipedalism, and brachiation. This holds true also for the langurs and colobus monkeys, which are designated semibrachiators, which means that they mainly move quadrupedally (usually with a "galloping gait" rather than walking) but also jump across gaps and occasionally swing by their arms. Although the categories are phrased in behavioral terms, their implications are also anatomic. Brachiation is the mode of locomotion for which the animal is specifically adapted; the anatomic correlates of brachiation are quite unmistakable and can be determined in fossil bones as much as in living animals. In some instances, it may well be that a particular anatomy is misleading. It has been argued that the anatomy of the great apes (including humans), for instance, is that of a brachiator, yet in fact they seldom brachiate (humans rarely and adult gorillas probably never). Watching gorillas, in particular, suggests that what appeared at first to be a locomotor adaptation may actually be a feeding one; the gorilla sits erect amid its food, reaching all around it to pull it in, and thereby saves an enormous amount of energy. The shortened lumbar spine (giving a lowered centre of gravity), broad chest, enhanced mobility of the shoulder joint, and flexible wrist may be related to this feeding style. The gibbons' specializations for brachiation may be derived from these same traits, rather than the other way about.

Changes in climate and geography during the evolutionary history of primates may also have led to structural atavisms in the anatomy of living primates. Many chimpanzees now living in woodland-savanna conditions in Africa, where the trees are widely spaced and generally unsuitable for the classic climbing style of forest-living chimpanzees, have adopted a largely ground-living life. Gorillas and chimpanzees are first and foremost knuckle walkers, but, given an environment like that of a zoo with a cage specially designed with lots of overhead bars and ropes, they will brachiate fairly frequently.

When the subject of primate arboreal locomotion is studied in evolutionary terms by using fossils, it becomes clear that locomotor categories are not discrete but constitute a continuum of change from a hind limb-dominated gait to a forelimb-dominated one. The best single indicator of gait, one that has the added advantage of being strictly quantitative, is the intermembral index. Briefly, the index is a ratio expressed as percentage of arm length to leg length; an index over 100 indicates relatively long arms. This provides a model by means of which the locomotion of an early primate can be inferred by determination of the intermembral index of the fossil skeleton. Animals do not necessarily fall discretely into categories. Species with indexes lying between those of clearly recognizable locomotor types represent transitional types, whose style of locomotion really does manifest features of both of the bracketing categories. Some lemurs have indexes that fall between 65 and 75, and their gait is a combination of vertical clinging and quadrupedalism. The South American spider monkeys (genus Ateles), whose index lies between 100 and 108, show a type of locomotion that contains the elements of both quadrupedalism and brachiation.

When the intermembral index is applied to fossil primates, it appears that the earliest primates living in the Eocene Epoch (56 million to 34 million years ago) must have moved about somewhat in the manner of modern vertical clingers and leapers. Quadrupedal gaits were well established during the Miocene Epoch (23 million to 5.3 million years ago) when the two major environmental types of quadrupedal gait— terrestrial and the arboreal—were established, with indexes in the region of 85–100 and 75–85, respectively. Brachiation, associated with a high intermembral index, was established as a way of arboreal life at the end of the Miocene, with the small hominid Oreopithecus from Italy. There is direct evidence of bipedalism's extending back four million years, and certain indirect evidence suggests that bipedalism might have evolved in a modified form up to a million years before that.

Bipedalism

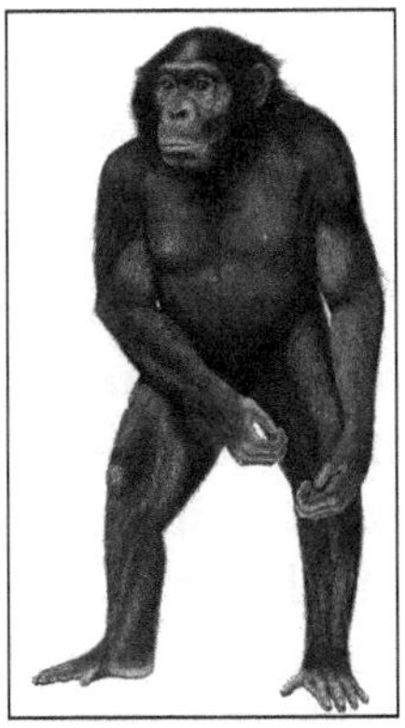

Bonobo, or pygmy chimpanzee (Pan paniscus), male.

Some degree of bipedal ability, of course, is a basic possession of the order Primates. All primates sit upright. Many stand upright without supporting their body weight by their arms, and some, especially the apes, actually walk upright for short periods. The view that the possession of uprightness is a solely human attribute is untenable; humans are merely the one species of the order that has exploited the potential of this ancestry to its extreme.

Chimpanzees, gorillas and gibbons, macaques, spider monkeys, capuchins, and others are all frequent bipedal walkers. To define humans categorically as "bipedal" is not enough; to describe them as habitually bipedal is nearer the truth, but habit as such does not leave its mark on fossil bones. Some more precise definition is needed. The human walk has been described as striding, a mode of locomotion defining a special pattern of behaviour and a special morphology. Striding, in a sense, is the quintessence of bipedalism; it is a means of traveling during which the energy output of the body is reduced to a physiological minimum by the smooth, undulating flow of the progression. It is a complex activity involving the joints and muscles of the whole body, and it is likely that the evolution of the human gait took place gradually over a period of 10 million years or so.

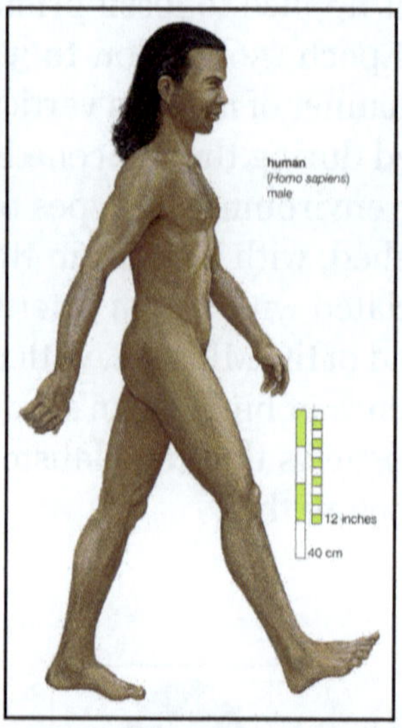

Human being (Homo sapiens), male.

The pattern of locomotion of human ancestors immediately preceding the acquisition of bipedalism has long been a matter of controversy, and the question has not yet been resolved. The evidence derived from anatomic, physiological, and biochemical studies for the close affinity of chimpanzees and humans, and the slightly less close affinity of gorillas, would suggest that humans evolved from a knuckle-walking ancestry. There have been claims that the wrist anatomy of australopithecines shows remnant knuckle-walking adaptations. The issue is still hotly debated, and some authorities continue to support a brachiation model for the ancestry of all the apes. Other authorities have proposed other solutions: semibrachiation, for example, and even a form of locomotion similar to that of tarsiers and other clingers and leapers. At the present time, there is insufficient information to elucidate the phylogeny of man's bipedal gait, except that it can be assumed to have involved a large measure of truncal uprightness.

Artist's rendering of Australopithecus afarensis, which lived from 3.8 to 2.9 million years ago.

Diet

The diet of primates is a factor of their ecology that, during their evolution, has clearly played an important role in their dispersion and adaptive radiation as well as in the development of the teeth, jaws, and digestive system. Diet is also closely related to locomotor pattern and to body size.

Aye-aye (Daubentonia madagascariensis) eating a coconut.

The principal food substances taken by primates may be divided into vegetable (fruits, flowers, leaves, nuts, barks, pith, seeds, grasses, stems, roots, and tubers) and animal (birds, birds' eggs, lizards, small rodents and bats, insects, frogs, and crustacea). The flesh of larger mammals (including primates) is not listed as an important item of nonhuman primate diet, with the sole exception of chimpanzees—it is taken by baboons in special circumstances that are not yet fully understood.

While diet is selective and specific to the order in many mammalian groups, among primates it is difficult to establish any hard and fast rules. Although there are decided preferences for certain food items, catholicity is more characteristic than specificity. Generally speaking, primates are omnivorous, as the physiology of their digestive system attests. Relatively few examples of dietary specialization are to be found. The so-called leaf-eating monkeys, a sobriquet that embraces the whole of the subfamily Colobinae, including colobus monkeys and langurs, are by no means exclusively leaf eaters and according to season include flowers, fruit, and seeds in their diet. The howler monkeys of the New World have a similar dietary preference.

Broadly, however, certain overall dietary preferences are discernible. The leaf-eating langurs have already been mentioned. The apes (other than the mountain gorilla) are substantially fruit eaters. Many of the smaller nocturnal primitive species such as galagos, dwarf lemurs, sportive lemurs, the aye-aye, and the slender loris are substantially insectivorous; the tarsier is probably the only primate that is exclusively carnivorous, feeding on insects, lizards, and snakes. The larger diurnal lemurs (e.g., typical lemurs, the sifaka, and the indri) are more vegetarian, including fruit, seeds, and leaves. It seems apparent that size, rather than activity rhythm, governs the nature of the primate diet. The small marmosets of the South American genus Callithrix have exclusively diurnal rhythms and are insectivorous and also eat gums, while the slightly larger, but equally diurnal, tamarins (genus Saguinus) are more omnivorous. An approximate cutoff point of 500 grams (Kay's threshold, after the primatologist Richard Kay, who first drew attention to it) has been proposed as an upper limit for species subsisting mainly on insects and a lower limit for those relying on leaves. The reason is that insects are small and hard to catch, and a large animal simply would not be able to catch enough to sustain it during its waking hours. The cellulose and hemicellulose components of leaves, on the other hand, require complex digestive processes, and a small animal would be unable to maintain a constant throughput. Fruit, as a dietary component, suffers from neither of these constraints.

Size in Evolutionary Perspective

In evolutionary terms, increase in size has probably played a large part in determining the direction of primate evolution. Early primates of about 50 million years ago were small forest-living creatures whose molar teeth bore high, pointed cusps but were neither as tall nor as pointed as those of their insectivore-like ancestors, whose molars were ideally adapted for cracking the hard chitinous exoskeletons of insects. This fact suggests that the reduction of the molar cusps was associated with the adoption of a fruit-eating habit. Although this has some validity as a generalization, it should not be taken too literally, as most primates include some insects in their diet and of course there are many almost exclusively insectivorous forms, which have nonetheless reduced the height and acuity of their molar cusps. Increasing body size, a trend that is clearly apparent throughout primate evolution, would have been associated with the adoption of supplementary sources of food. An increase in size and the gradual addition of bulk foods to the diet would in turn have affected the habitat and the pattern of locomotion of primates. Suitable adaptations in this case would have been the facility to climb, leap, and balance in trees.

It is noteworthy that, during evolution, the development of a prehensile foot preceded that of a prehensile hand. Vertical-clinging primates such as the tarsiers or small, squirrel-like quadrupeds such as the marmosets—all of which have prehensile feet but not completely prehensile hands—by remaining or becoming small, have avoided the evolutionary pressures that have impinged on larger primates. A large arboreal primate without entirely prehensile hands is at a considerable disadvantage in moving about in the canopy of trees, but a small one suffers little disadvantage. Amid the large and firm branches, size is no particular hazard, but at the periphery of the crown, where the fruit is most abundant and the branches are slender and flexible, the risk of falling is increased. It is therefore likely that the combination of an increase in body size associated with the inevitable shift toward a bulk diet led first to the evolution of a grasping hand, then to the appearance of a prehensile hand, and finally to an opposable thumb. Four prehensile extremities are obviously more effective than two in defying gravity.

Such adaptations of the forelimbs would have had the effect of equalizing the role of the limbs. The limbs of vertical clingers are functionally disparate, the lower pair being dominantly propulsive and the upper secondary and purely supportive. The limbs of quadrupeds, however, are more homogeneous, both pairs having a propulsive function during running. Thus, it would seem that the transition in locomotor grade between vertical clinging and leaping and quadrupedalism came about as an adaptation to increased body size. Size, diet, ecology, locomotion, and anatomic structure provide a constellation of causes and effects that are critical factors in the evolution of the primates.

Forest and Savanna

The chief physiognomic features of rainforests, the ancestral home of the order Primates and the principal habitat of nonhuman primates today, are the evergreen broad-leaved trees that collectively form a closed canopy, so opaque to sunlight that the forest floor is in perpetual twilight. Epiphytes and thick-stemmed lianas drape the trees, linking one crown to another and providing aerial pathways for monkeys to pass from tree to tree through a continuum of interlacing branches, a three-dimensional maze that provides home, restaurant, shopping districts, and highways for primates. Three strata of rainforests are broadly distinguishable: an understory, a middle story, and an upper story. The understory, consisting of shrubs and saplings, is often "closed," the crowns of the constituent trees overlapping one another to form a dense continuous horizontal layer. The middle story is characterized by trees that are in lateral contact but do not overlap; the highest story, by tall trees, some 50 metres (about 165 feet) or more, that form a discontinuous layer of umbrella-shaped crowns. The occasional "emergent" forest giant may tower above the highest layer of the canopy. There is some evidence, much of it conflicting, that some zonation of forest primates occurs within the forest canopy. The stratification of forest is extremely variable; the number of layers tends to diminish from three to two in secondary forest, dry deciduous forest, and montane forest and from two to one as temperate zone, tropical woodland, or montane woodland supervenes.

Gibbons (family Hylobatidae).

Tropical grasslands, or savannas, are also the homes of primates in Africa and Asia; no savanna-living primates exist in South America. Tropical grasslands comprise a mixture of trees and grasses, the proportion of trees to grass varying directly with the rainfall. Areas of high seasonal rainfall support single-story woodlands of tall trees, while lush grasses form the ground vegetation; but, where rainfall is both seasonal and low, the trees consist of stubby xerophilous (dry-loving) shrubs and short, tussocky grasses. The principal primates of the savanna are the ground-living species: in Africa, the vervets, baboons, and patas monkey; and in Asia, the macaques and the Hanumān langur.

Tropical montane forests or tropical rainforests at high altitude also abound in primates in Africa, Asia, and South America. In equatorial Africa, certain primate species have colonized the montane-savanna regions, or moorlands, where the rugged mountainous terrain and seasonal food scarcity support herds of geladas and hamadryas baboons. These high mountaineers of Africa have no ecological counterparts in Asia or South America.

Form and Function

General Structure

The basis of the success of the order Primates is the relatively unspecialized nature of their structure and the highly specialized plasticity of their behaviour. This combination has permitted the primates throughout their evolutionary history to exploit the wide variety of novel ecological opportunities that have come their way. Although there are a few highly specialized species among the lower primates (the aye-aye, the tarsier, the potto, and the lorises, for instance), the higher primates, collectively known as the anthropoids, are extremely conservative in their structure; morphologically speaking, they have maintained a position in the evolutionary midstream and have avoided the potential stagnation of specialized life near the banks. Specialization is not always a liability; in times of environmental stability, the specialized animal enjoys many advantages, but, in a rapidly changing world, it is the less-specialized animals that are more likely to survive and flourish. The plasticity of primate behaviour is largely a function of the brain. The primate brain is distinguished by its relatively large size compared with the size of the body as a whole; it is also notable for the complexity and elaboration of the cerebral cortex, the function of which is to receive, analyze, and synthesize the incoming impulses from the sense organs and to convert them into appropriate motor actions, which in turn constitute behaviour.

Guereza, or Abyssinian, black-and-white colobus (Colobus guereza).

Primates are essentially arboreal animals whose limbs are adapted for climbing, leaping, and running in trees. Active arboreal life requires the mechanical assistance of a long tail and sensitive, grasping hands and feet with opposable thumbs and big toes to aid in climbing and to ensure stability on slender branches high above the ground. Active arboreal locomotion also requires a much more accurate judgment of distances than life on the ground; this is facilitated by the development of stereoscopic vision, the anatomic basis of visual judgments in depth. The forward-facing eyes of primates are adaptations for this type of visual precision. A highly developed sense of smell is not nearly as important for animals leading an arboreal life as it is for those on the ground. Many primates thus have a much-reduced olfactory mechanism; noses are shorter, and the nasal concha (scroll bones) of the nose are reduced in number and complexity compared with most nonprimate mammals—although it should not be overlooked that many lemurs and New World monkeys do enjoy a rich olfactory world, especially in the social sphere.

Above all, the principal evolutionary trend of primates has been the elaboration of the brain, particularly of that portion of the cerebral hemispheres known as the neopallium or neocortex. A neocortex is characteristic of higher vertebrates, such as mammals, which operate under the control of multiple sources of sensory input. In many mammals, the olfactory system dominates the senses, and the cerebral hemispheres consist largely of palaeocortex—the "smell brain"—of lower vertebrates. The arboreal habit of primates has led to a dethronement of the olfactory sense and the accession of a tactile, visually dominant sensory system. This evolutionary trend has resulted in the dramatic expansion and differentiation of the neocortex.

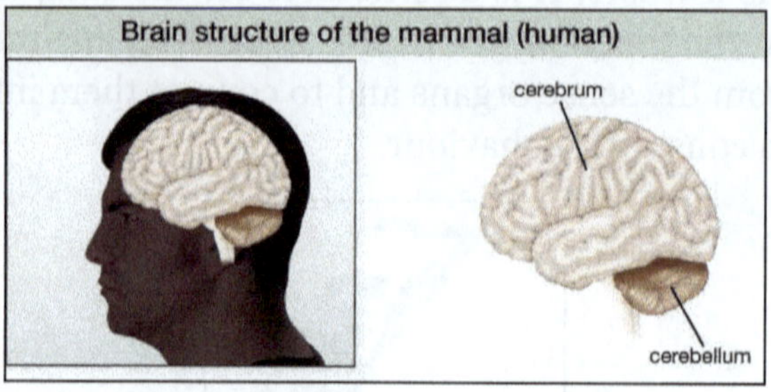

In the brains of primates such as humans, the cerebrum has grown into the largest region of the brain and has a characteristic lobed appearance.

Vertebral Column and Posture

All primates retain collarbones (lost in many mammalian groups), a separate radius and ulna in the forearm, and a separate tibia and fibula in the lower leg. The single exception to this among living primates is the tarsier, in which the fibula becomes fused to the tibia.

The primate vertebral column shows a basic mammalian pattern of components, including an "anticlinal" vertebra situated in the mid-thoracic (upper-back) region of the spinal column and marking the transition between the forelimb and hind limb segments. In a galloping greyhound, the anticlinal vertebra is at the apex of the acute curve of the back. An anticlinal vertebra is characteristic of all quadrupeds and is seen in all primate families except the apes, whose posture is upright or semiupright. The evolutionary trend in the vertebral column is toward shortening the lumbar, sacral, and caudal regions. Extreme shortening in the lumbar region, with complete loss of the caudal (tail) vertebrae, occurs in gibbons, the great apes, and humans. Prehensility of the tail is a specialization of certain New World monkeys, but it appears also for a brief period in the infants of many Old World monkey groups, in which it provides an important mechanical aid to survival.

Hands and Feet

With three exceptions, all primates have retained five digits on hand and foot. The exceptions are the spider monkeys and the so-called woolly spider monkey of South America and the colobus monkeys of Africa, which have lost or reduced the thumb. This appears to be an adaptation for locomotion, the rationale for which is not fully understood at present.

Sooty mangabey (Cercocebus atys).

All, though to different degrees, possess prehensile (grasping) hands and all (except humans) prehensile feet. The hands of catarrhines show a greater range of precise manipulative activity than those of other primates. Lemurs, for example, lack the functional duality of the hands of most apes and Old World monkeys (catarrhines). Duality in hand function has been described in terms of precision and power grips. The power grip of lemurs and lorises is very well developed, but the precision grip is lacking. The New World monkeys show a considerable advance over primitive primates in tactile sensitivity, but they possess less functionally effective hands in prehensile terms than Old World monkeys.

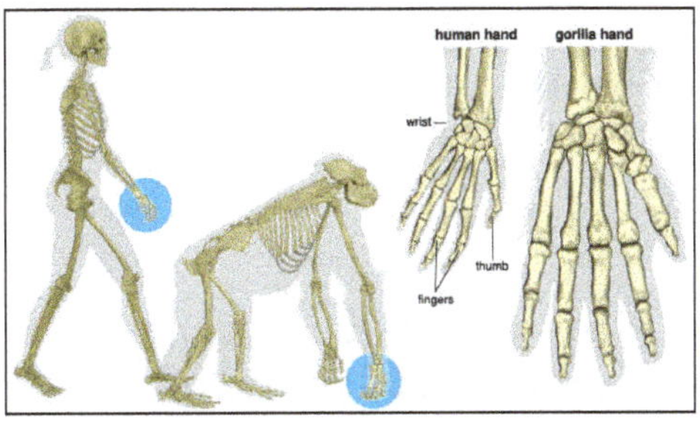

Human and gorilla hands: The hand structure of a human being (left) and of a gorilla (right).

Orangutan (Pongo pygmaeus), female.

The critical component of the prehensile hand in terms of skilled manipulation is the opposable thumb—a thumb, that is to say, that is capable of being moved freely and independently. The movement of opposition is a rotary movement in which the thumb, swinging about its own axis, comes to face the lower surface of the tips of the fingers. The opposable thumb is the basis of the precision grip that, though present to some extent in all primates, is particularly highly developed in man. Opposability is present to some degree in most primates but varies considerably in its functional effectiveness as an instrument of fine manipulation. Humans and baboons are preeminent in this respect. The apes, having short thumbs and long fingers, are handicapped in relation to delicate manual dexterity but are adept in the coarser elements of hand use, particularly in relation to tree climbing.

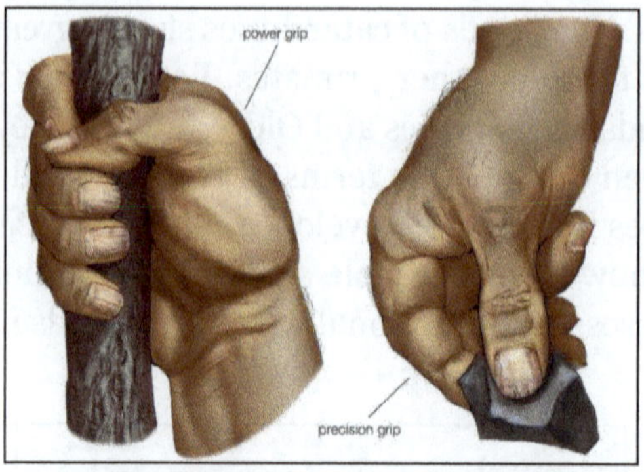

A fully opposable thumb gives the human hand its unique power grip (left) and precision grip (right).

Teeth

A dentition with different kinds of teeth (heterodonty)—incisors, canines, and cheek teeth—is characteristic of all primates and indeed of mammals generally. Heterodonty is a primitive characteristic, and primates have evolved less far from the original pattern than most mammals. The principal changes are a reduction in the number of teeth and an elaboration of the cusp pattern of the molars.

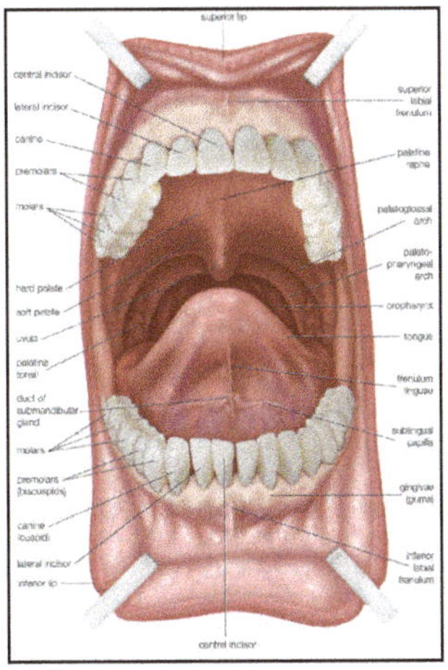

Anterior view of the oral cavity.

The dental formula of primitive placental mammals is assumed to have been 5 . 1 . 4 . 3 / 5 . 1 . 4 . 3 = 44 teeth (the numbers being the numbers respectively of pairs of incisors, canines, premolars, and molars in the upper and lower jaws). No living primate has retained more than two incisors in the upper jaw. The incisors are subject to considerable variation in strepsirrhines. Characteristically, the upper incisors are peglike, one or the other pair often being absent; in the lower jaw, the incisors show a peculiar conformation that has been likened structurally and functionally to a comb. This dental comb is composed of the lower canines and lower incisors compressed from side to side and slanted forward; the most specialized dental combs—seen, for example, in the fork-crowned lemur (genus Phaner) and the needle-clawed galago (genus Euoticus)—are used for scraping exudates off bark, but other species use the structure for piercing fruit, for nipping off leaves, and for grooming the fur. Canines are present throughout the order but show remarkable variation in size, shape, projection, and function. Characteristically, the teeth of Old World monkeys have a function in the maintenance of social order within the group as well as an overtly offensive role; their function as organs of digestion is relatively unimportant. They are large and subject to sexual dimorphism, being larger in males than females. Great apes have smaller canines than Old World monkeys, though still sexually dimorphic; human canines are smaller still, and there is no size difference between the sexes.

The trend in the evolution of the cheek teeth has been to increase the number of cusps and reduce the number of teeth. Both molars and premolars show this tendency. No living primate has four premolars; primitive primates, tarsiers, and New World monkeys have retained three on each side of each jaw, but in the apes and Old World monkeys, there are only two premolars. The primitive premolars are uniform in shape and are unicuspid, but in primates the most posterior premolar tends to evolve either one or two extra cusps (molarization), an adaptation that extends the cheek-tooth row for a herbivorous diet. In species with large upper canines, the most anterior lower premolar assumes a peculiar shape known as sectorial, functioning as a hone for the scythelike canine. In humans, whose canines are small and unremarkable, the first and second premolars are identical in shape and two-cusped.

The trend in the morphology of the molars has been to increase the primitive three cusps to four or five, the less-insectivorous species having four cusps on the molar crown in the upper jaw and five cusps on the lower. A tendency in smaller New World monkeys has been to reduce the molar series from three to two in both jaws.

Snouts, Muzzles, and Noses

The reduction of the snout in primates is a correlate of the diminution of the sense of smell, or olfaction. To a great extent, visual acuity and manual dexterity have replaced the sensitive, inquiring nose found in so many nonprimate mammals. A marked reduction in the complexity of the nasal concha ("scroll" bones of the nose), the richness of the innervation of the olfactory mucous membrane, and the sensitivity of the moist tip of the nose—the rhinarium—are associated with the reduction in length of the primate snout. Still, although the trend in primate evolution is toward a dethronement of the primacy of the sense of smell, there are still some good snouts to be seen in those lower primates that retain a naked moist rhinarium attached to the upper lip.

Male hamadryas (Papio hamadryas).

Lemurs, lorises, tarsiers, and New World monkeys depend for many aspects of their social and reproductive behaviour on olfactory signals, by means of special scent glands distributed in different regions of the body but congregated principally in the anal and perineal regions (in lemurs and lorises) or in the sternal region (in New World monkeys and tarsiers). Marking behaviour, the placing of scent at various points in the environment, is a prominent feature of the repertoire of communication in these primates. Marking behaviour ceases to be of much importance in the Old World monkeys and apes, with some exceptions such as mandrills, siamangs, concolor gibbons, and even orangutans among the hominoids. All of these primates possess sternal glands, but in all of them the structures by which olfactory signals are given and received are diminished. But all higher primates, including humans, sniff at unfamiliar items of food before placing them in the mouth.

The shape of the nose of higher primates is one of the most reliable means of distinguishing Old World monkeys from New World monkeys at a glance. In New World monkeys (the Platyrrhini, meaning "flat nosed"), the nose is broad, and the nostrils are set wide apart, well separated by a broad septum, and point sideways. In the apes and Old World monkeys (the Catarrhini, meaning "downward nosed"), the nostrils are set close together, point forward or downward, and are separated by a very narrow septum.

Pale-headed saki (Pithecia pithecia).

Male drill (Mandrillus leucophaeus).

Some Old World monkeys—particularly those that have adopted a ground-living way of life, such as baboons and mandrills, of the subfamily Cercopithecinae—appear to have readopted a long snout during their evolution. This structure, however, is not primarily olfactory in function but seems, rather, to be allometric, more closely related to the large size of the jaws and the prominence of the canine teeth; it should be considered a dental muzzle rather than an olfactory one.

Sensory Reception and Brain

Among mammals in general, the olfactory system is the primary receptor for environmental information; consequently, the brain of most mammals is dominated by the olfactory centres. In primates the sense of smell is considerably less important than the well-developed visual system and highly refined sense of touch. The primate brain is enlarged in the specific areas concerned with vision (occipital lobes) and touch (parietal lobes) and thus takes a characteristic shape throughout the higher primates.

Cotton-top tamarin

Cotton-top tamarin (Saguinus oedipus) used in an experiment to determine if music containing calls from others of the same species affected behaviour.

Touch

The skin of the primate hand is well adapted for tactile discrimination. Meissner's corpuscles, the principal receptors for touch in hairless skin, are best developed in apes and humans, but they can be found in all primates. Structurally correlated with a high level of tactile sensitivity are certain anatomic features of the skin of the hands and feet, such as the absence of pads on the palms and soles and the presence of a finely ridged pattern of skin corrugations known as dermatoglyphics (the basis for fingerprints).

Enormous eyes and padded digits are adaptations that evolved in the tarsier (Tarsius), a nocturnal, arboreal primate of the rainforests of Southeast Asia.

Eyes and Vision

The evolutionary trend toward frontality of the eyes has not proceeded as far in most lemurs as in lorises and more advanced primates. In primitive mammals the central axes of the two bony orbits are 140° apart. In lemurs this angle is considerably less, 60°–70°, and in the apes and monkeys and in the slender loris (genus Loris), the divergence has been reduced to 20°. It should be noted that the axes of the eyeballs (as distinct from the bony orbits) in apes and monkeys are, in fact, parallel.

The moustached monkey (Cercopithecus cephus) is a species of guenon.

Male gorilla (Gorilla gorilla).

Colour vision is of considerable advantage to arboreal animals living on fruits and insects. Most mammals have both rod and cone receptors in their retinas, and almost all primates have at least two kinds of cones, a short-wavelength (blue) type and a medium–long-wavelength (red-green) type. All, therefore, seem to have well-developed colour vision, the exceptions so far known being some of the nocturnal species: durukulis of South America, the tarsiers, and at least some of the galagos. Catarrhines and howler monkeys have separate red- and green-responding cones, determined by closely linked loci on the X chromosome. In most other investigated platyrrhines (New World monkeys), red and green are determined by alleles at a single locus, again on the X chromosome; thus, males are always dichromatic, whereas females may be either dichromatic (if homozygous) or trichromatic (if heterozygous). It is suggested that, during the evolution of catarrhines, the red-green locus duplicated, one of the daughter loci fixing the red gene and the other the green.

Nervous Integration

The elaboration of touch and vision supplements the senses of smell, hearing, and taste, providing the primate with a sensory armament of great range and flexibility. The primate central nervous system is sufficiently refined to deal with the elaborate bombardment of environmental information reaching it. Association areas provide connections between the input and output centres of the brain—the motor and sensory cortex. Association areas are the memory banks where the memory of past experience is encoded in the infinitely complicated plexiform arrangement of the neurons, the brain cells, and their processes. All sensory impulses reaching the cortical centres of the central nervous system are routed through the association areas for conditioning, as it were, before reaching the effector side or output side where the appropriate response is initiated in the cells of the motor cortex. The more highly developed the association areas of the brain are, the more specific and appropriate is the behaviour and the more versatile is the animal in facing environmental demands.

The Brain

The principal evolutionary trend in brain development has been toward elaboration. The neocortex of higher primates possesses highly developed associative functions, an aptitude for receiving, analyzing, and synthesizing the sensory input from visual, olfactory, auditory, gustatory, and tactile receptors and converting them into the appropriate motor responses.

The brain of monkeys and apes is larger, both absolutely and relatively, than that of lemurs, lorises, or tarsiers. For instance, the weight of the simplest anthropoid brain, that of a marmoset, is three times greater than the brain weight of a bush baby of comparative size. This quantitative increase is attributable in part to the elaboration of the regions of the neocortex concerned with tactile and visual sensitivity and in part to the elaboration of the intrinsic pathways connecting one part of the brain with another. The large brain of humans is attributable not so much to an increased nerve cell content as to an increase in the size of the nerve cells and to a greater complexity of the connections linking one cell to another.

The external form of the anthropoid cerebral cortex is characterized by a complicated pattern of folds and fissures (sulci and gyri) in the brain surface. The fissural pattern is seen in its simplest form in the marmosets, but in the larger New World monkeys (capuchins, for instance), the cerebrum is richly convoluted. Gyri and sulci are well marked in Old World monkeys and in the apes, the complexity of the pattern closely approximating the tortuous mazelike pattern seen in humans.

Reproductive System

Male and Female Genitalia

The functions of the individual organs of reproductive systems are fairly uniform throughout the primates, but, in spite of this physiological homology, there is a remarkable degree of variation in minor detail of organs between groups—particularly in the external genitalia, which, by their variation, provide a morphological basis for the reproductive isolation of the species. There could be no more effective barrier to mating between different species than incompatibility of the male and female sex organs.

Among the characteristics of the primate order as listed by the 19th-century zoologist George Mivart, the penis is described as "pendulous" and the testes as "scrotal." In contrast to most other mammals (bats being the principal exception), the primate penis is not attached to the abdominal wall but hangs free. The testes, with a few exceptions among the lemurs, in which they are withdrawn seasonally, lie permanently in the scrotal sac, to which they migrate from their intra-abdominal position some time before birth (in humans) or after birth (in nonhuman primates). In all primates except modern humans, tarsiers, and some South American monkeys, the penis contains a small bone called the baculum, a typically mammalian character. The uterus of female primates shows all grades of transition between the two-horned uterus, typical of most mammals, to the single-chambered uterus of the higher primates and humans.

Variations between primate taxa are demonstrated most strikingly by the glans penis, scrotum, and perineum of the male and by the clitoris and labial folds of the female vulva. In the clitoris, there is in most primates a small bone, the baubellum, homologous with the baculum of the penis. The length and form of the clitoris, which when elongated mimics the penis (as in spider monkeys, for instance), are a potent source of confusion in determining the sex of certain New World primates. The coloration of the male scrotum in forest-living primates, particularly of the guenon (genus Cercopithecus) and in drills and mandrills (genus Mandrillus), shows an infinite range of variations and provides a species-recognition signal of considerable effectiveness.

The external appearance of the genitalia undergoes seasonal variation in a number of primates. In the male, swellings of the testes and colour changes of the scrotum occur, and, in the female, swelling and coloration of the vulva and perineal region herald ovulation, sometimes most obtrusively. Turgidity and excessive vascularity of the tissues of the perineum are probably characteristic of all mammals, but there are certain primate species in which this engorgement reaches monstrous proportions, notably baboons, mangabeys, some macaques, and chimpanzees. Regions other than the primary sex organs may also be affected by hormones circulating at certain periods of the reproductive cycle. For instance, in the gelada, the skin on the front of the female chest, which normally bears a string of caruncles resembling the beads of a necklace, becomes engorged and brightly coloured. A German zoologist, Wolfgang Wickler, has suggested that this is a form of sexual mimicry, the chest mimicking the perineal region. The observation that geladas spend many hours a day feeding in a sitting posture provides a feasible, Darwinian explanation of this curious physiological adaptation.

Placenta

The placenta, the defining characteristic of all eutherian mammals, is a vascular structure that permits physiological interchange of blood and body fluids between the mother and the fetus and the breakdown products of the fetal metabolism; it also provides a two-way barrier preventing the passage of some, but not all, noxious substances and organisms such as bacteria and viruses from one individual to the other and is the source of hormones such as estrogens.

The placenta is a flat, discoid-shaped "cake" in humans, some of the other monkeys and apes, and the tarsiers. In many monkeys, it is bidiscoidal, having two linked portions. The placenta is intimately attached on its outer surface to the endometrium, the lining of the uterus, by fingerlike processes (villi) that embed themselves in the endometrium, where complete vascular connections between the two circulations are achieved. The connection between fetal and maternal circulations appears as two distinct types among primates, a distinction that is believed to have had an important effect on the evolution of the order. In the first type (epitheliochorial), found in the lemurs and lorises, several cellular layers separate the maternal and fetal bloodstreams and thus limit the passage of molecules of serum proteins. In the second type (hemochorial), found in tarsiers, monkeys, and apes, the relationship is much more intimate, there being no cell layers separating the two circulations so that serum proteins can easily pass. In haplorrhines the endometrium becomes highly vascularised about two weeks after ovulation, in preparation for the possible implantation of a zygote; if this does not occur, it is shed via menstruation. The placenta is shed at birth in all primates and, except rarely among humans, is eaten by the mother.

References

- Mammal, animal: britannica.com, Retrieved 15 March, 2019
- Prototheria, accounts: animaldiversity.org, Retrieved 5 May, 2019
- Metatheria, accounts: animaldiversity.org, Retrieved 17 February, 2019
- Evolutionary-history, Eutheria: newworldencyclopedia.org, Retrieved 7 July, 2019
- Afrosoricida: newworldencyclopedia.org, Retrieved 23 April, 2019
- Afrosoricida, eutheria: mammalsrus.com, Retrieved 27 August, 2019
- Macroscelidea-sengis, environment: encyclopedia.com, Retrieved 17 March, 2019
- Primate-mammal, animal: britannica.com, Retrieved 3 January, 2019

4

Fundamentals of Ornithology

The branch of zoology which focuses on the study of animals which belong to the class Aves, also called birds, is termed as ornithology. Birds are broadly divided into two groups, ratites and non-ratites. The chapter closely examines the key aspects of these types of birds to provide an extensive understanding of the subject.

Ornithology is a branch of Zoology or Biology concerned with the scientific study of birds, creatures belonging to class Aves.

The science of Ornithology includes the study of classification of birds, evolution, body structure, habits, song, flight and breeding behavior. Birds are generally divided into two major groups. Ostriches, Emus, and Kiwis are ratites, flightless birds that lack keel-like extension on the breast bone. The vast majority of the birds, however, are non-ratites and possess the keel to which the powerful muscles that power the wings are attached. There are over 9200 extant species of birds divided into 30 orders & 174 families. Their ability to fly has allowed them to colonize every land mass on earth including the Arctic and Antarctic and some, like the Penguins have adapted an aquatic lifestyle coming to shore only to breed and raise young. Fossil evidence of the ancestors of modern birds indicates that they were contemporaneous with dinosaurs.

Types of Ornithology Study

The field of ornithology is vast, but almost all aspects of its study can be broken down into one of three fields: collection, field-based research, and laboratory-based research:

- Collection: refers to the more traditional biological approach to the study of birds, through the collection of both live and deceased specimens. The specimens are analyzed in terms of their body systems, anatomy, physiology, and even their DNA.
- Field-based research: involves scientists working out in the field, noting population counts, taking bird song recordings, and sometimes collecting (either permanently or temporarily) live specimens for more direct observation.

Field-based ornithologists also make indirect observations, which include studying the nesting locations, nesting materials, or food resources available to the birds living in a particular location. They aren't direct observations of a bird, but instead are observations of conditions or settings that a bird would live in or interact with.

- Laboratory-based research: can include analyzing the results of field-based studies or analyzing live birds for studies relating to animal intelligence, bird-to-bird and bird-to-human interactions, and medical treatments.

Aves

Aves (Birds) are bipedal, warm-blooded, oviparous (egg-laying) vertebrates characterized primarily by feathers, forelimbs modified as wings, and a bony beak without teeth. Common characteristics of birds include the laying of hard-shelled eggs, high metabolic rate, and a light but strong skeleton, generally with hollow bones.

The only animals covered with feathers, most birds are characterized by flight. Many birds migrate long distances to utilize optimum habitats (e.g., Arctic tern), while others spend almost all their time at sea (e.g., the wandering albatross). Some, such as frigatebirds, stay aloft for days at a time, even sleeping on the wing. The fastest living animal is a bird, the peregrine falcon (Falco peregrinus), which has been observed to reach a flying speed of at least 124 mph (200 km/h) and perhaps 350 km/h (217 mph) during a dive. On the other hand, the ratites are flightless, and several other species, particularly on islands, have also lost this ability. Flightless birds include the penguin, ostrich, kiwi, and the extinct dodo.

Birds range in size from the tiny hummingbirds to the huge ostrich and emu. Depending on taxonomic viewpoint, there are about 8,800 to 10,200 living bird species in the world, making them the most diverse class of terrestrial vertebrates.

The path of humans and birds have intersected in many ways throughout human culture, whether directly through hunting and domestication, or as powerful symbols in literature, arts, and religion. The dove has been used as a symbol of peace and love, as well as the Holy Spirit in Christianity; cranes have symbolized long life, and owls wisdom. The majority of birds are monogamous and many species mate for life, reflecting the higher aspirations of the human spirit and religion.

Birds are a very differentiated class, with some feeding on nectar, plants, seeds, insects, rodents, fish, carrion, or other birds. Some birds even feed on parasites on black rhinoceroses, or within the mouths of crocodiles. Most birds are diurnal, or active during the day. Some birds, such as the owls and nightjars, are nocturnal or crepuscular (active during twilight hours). Many birds seasonally migrate long distances, improving opportunities for food, better climate, and breeding.

Birds are among the most extensively studied of all animal groups. Hundreds of academic journals and thousands of scientists are devoted to bird research, while amateur enthusiasts (called birdwatchers or birders) probably number in the millions.

Bird Classification

Birds form a class, whose scientific name is Aves. Modern birds are classified in the subclass Neornithes, which is further separated into two superorders, the Paleognathae (mostly flightless birds), and the wildly diverse Neognathae, containing all other birds. There is also an extinct subclass, Archaeornithes.

The paleognaths include the tinamous (found only in Central and South America) and the ratites. The ratites are large flightless birds, and include ostriches, cassowaries, kiwis, and emus. Some scientists suspect that the ratites represent an artificial grouping of birds which have independently lost the ability to fly, while others contend that the ratites never had the ability to fly and are more directly related to the dinosaurs than other modern birds.

The classification of birds is a contentious issue. Sibley and Ahlquist's Phylogeny and Classification of Birds is an often cited work on the classification of birds, but is also frequently debated and constantly revised. Evidence has been accumulated that the modern bird orders constitute accurate taxa. However, scientists are not in agreement as to the relationships between the orders; evidence from modern bird anatomy, fossils and DNA have all been brought to bear on the problem but no strong consensus has emerged. Some postulated the Galloanseri as the basal divergence from the remaining Neognathes. The Galloanseri contain the Anseriformes (ducks, geese and swans), and the Galliformes (the pheasants, grouse, and their allies). While there is general consensus that the orders Anseriformes and Galliformes represent sister-groups, there remain opposing points of view.

Species and Habitats

The estimated number of living bird species is often put around ten thousand, but ranges from 8,800–10,200. This compares to about 4,800 species of mammals and 7,150 species of reptiles.

Birds have adapted to virtually every part of the globe and a wide diversity of habitats: forests, grasslands, cities, mountains, deserts, wetlands, and oceans. Even the snowy owl (Bubo scandiacus) is adapted to live north of the Arctic Circle, where the temperature can be below -46°C during the winter, and penguins thrive in the Antarctica. The albatrosses spend much of their time soaring over the open ocean, and the emperor penguin (Aptenodytes forsterican) plunges to depths of more than 250 meters (850 feet) searching for food.

However, birds have generally been adapted to terrestrial land—only two percent of all birds are aquatic. Most of these, about 150 species, are duck like birds, as well as a small number of grebes, loons, penguins and auks. Additionally, only a few birds are leaf-eaters, with a notable one being the hoatzin (Ophisthocomus hoazin).

Anatomy and Physiology

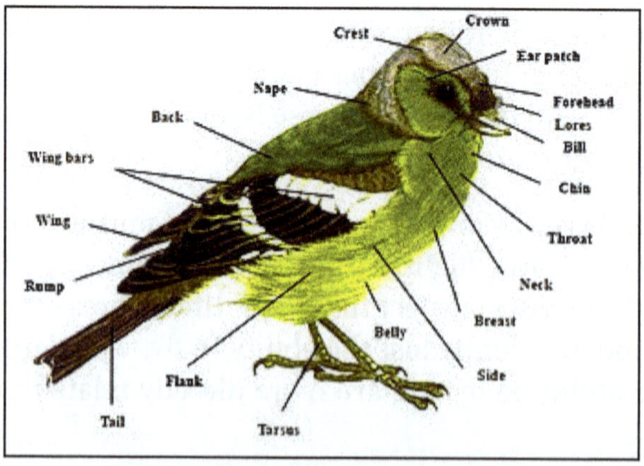

Anatomy of a typical bird.

Feathers

A distinguishing feature of birds is feathers. Other animals have wings and can fly, such as bats, but only birds are covered with feathers. Feathers are among the most complex structural organs found in vertebrates. These integumentary appendages are formed by controlled proliferation of cells in the epidermis (outer skin layer) from a protein called keratin. These fibrous structural proteins form the hard, but non-mineralized structures found in bird feathers, beaks, and claws, as well as reptilian claws, shells, and scales, and mammalian hair (including wool), nails, claws, horns, and hooves of mammals. The β-keratins used by birds and reptiles are even tougher than the α-keratins of mammals. Feathers insulate birds from water and cold temperatures. Individual feathers in the wings and tail play important roles in controlling flight. Color patterns serve as camouflage against predators for birds in their habitats, and by predators looking for a meal. Striking differences in feather patterns and colors are part of the sexual dimorphism of many bird species and are particularly important in selection of mating pairs. The blue and green colors in most parrots are attributed to a texture effect in microscopic portions of the feather itself, rather than pigment.

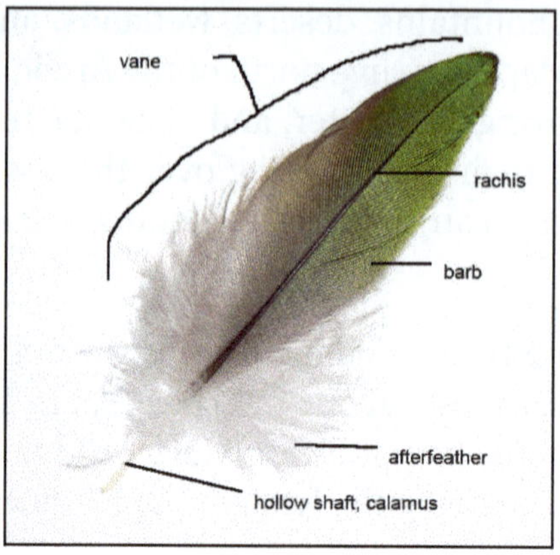

Parts of a contour feather.

Skeleton

Birds also have skeletons possessing unique characteristics.

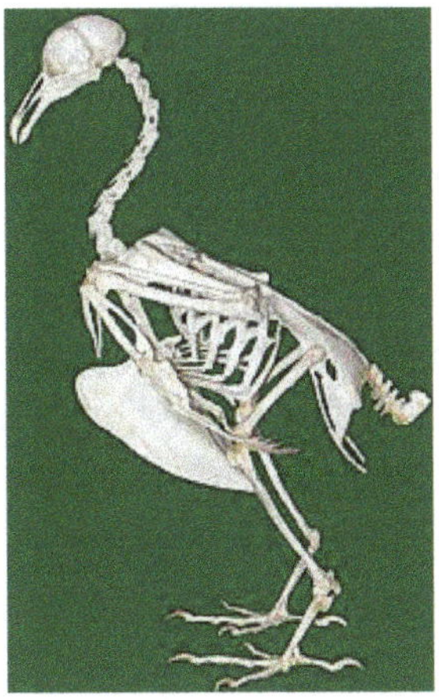

The skeleton of a dove.

The avian skeleton is highly adapted to these animals' capacity for flight. It is extremely lightweight, but strong enough to withstand the stresses that a bird experiences when taking off, flying and landing. One of the adaptations that make this possible is the fusing of bones that are separate in mammals into single ossifications. Because of this, birds usually have a smaller number of bones than mammals or reptiles, and the frame has more strength.

Birds have many bones that are hollow, with criss-crossing struts or trusses for structural strength. (Some flightless birds like penguins have only solid bones, however). The number of hollow bones varies from species to species, though large gliding and soaring birds tend to have the most.

Birds also have more cervical vertebrae than many other animals; most have a highly flexible neck that consists of 13-25 vertebrae. Although the neck is flexible, most of the skeleton is rather stiff. Birds are the only vertebrate animals to have a fused collarbone (the furcula or wishbone) or a keeled breastbone (or sternum). The heaviest and strongest bone is the coracoid.

Although feathers are light, a bird's plumage weighs two or three times more than its skeleton, since many bones are hollow and contain air sacks. The lack of teeth also lightens the frame.

The pectoral (breast) muscles of birds are the largest, and may account for 20% of a bird's weight.

Digestive System

A bird's digestive system extends from the mouth, through the pharynx, into the esophagus, into the stomach, through the small intestine and the large intestine, and out the cloacal opening. Some birds have a crop, which is a storage pouch that is part of the esophagus and can allow birds to feed and store food until later digestion. Most birds have a stomach made of two parts, the first part, or proventriculus, which secretes digestive fluids, and the second, the gizzard.

Lacking teeth, birds use their digestive system to grind and pulverize food. The ventriculus or gizzard is composed of four muscular bands that act to rotate and crush food by shifting the food from one area to the next within the gizzard. Depending on the species, the gizzard may contain small pieces of grit or stone that the bird has swallowed to aid in the grinding process of digestion. For birds in captivity, only certain species of birds require grit in their diet for digestion. The use of gizzard stones appears to be a similarity between birds and dinosaurs, based on trace fossils of gizzard stones called gastroliths.

Respiration

The high metabolism of birds necessitates an effective system for acquiring a great deal of oxygen.

Birds ventilate their lungs by means of crosscurrent flow: the air flows at a 90° angle to the flow of blood in the lungs' capillaries. In addition to the lungs themselves, birds have posterior and anterior air sacs (typically nine), which control airflow through the lungs, but do not play a direct role in gas exchange. There are three distinct sets of organs involved in respiration:

- The anterior air sacs (interclavicular, cervicals, and anterior thoracics),
- The lungs, and
- The posterior air sacs (posterior thoracics and abdominals).

It takes bird two full breaths to completely cycle the air from each inhalation through its lungs and out again. Air flows through the air sacs and lungs as follows:

- First inhalation: air flows through the trachea and bronchi into the posterior air sacs.
- First exhalation: air flows from the posterior air sacs to the lungs.
- Second inhalation: air flows from the lungs to the anterior air sacs.
- Second exhalation: air flows from the anterior sacs back through the trachea and out of the body.

Since during inhalation and exhalation fresh air flows through the lungs in only one direction, there is no mixing of oxygen rich air and carbon dioxide rich air within the lungs as in mammals. Thus the partial pressure of oxygen in a bird's lungs is the same as the environment, and so birds have more efficient gas-exchange of both oxygen and carbon dioxide than do mammals.

Avian lungs do not have alveoli, as mammalian lungs do, but instead contain millions of tiny passages known as parabronchi, connected at either ends by the dorsobronchi and ventrobronchi. Air flows through the honeycombed walls of the parabronchi and into air capillaries, where oxygen and carbon dioxide are traded with cross-flowing blood capillaries by diffusion.

Circulatory, Nervous and Metabolic Systems

Birds have four-chambered hearts, and a circulatory system with blood vessels. Birds' high metabolism requires rapid circulation because of the quick build up of waste products, and need for oxygen.

In general, birds generally have exceptional vision, well-developed hearing, and a poor sense of smell. As a group, their vision is the best of all animals. Large eyes, often heavier than their brain, offer strong visual acuity and color perception. However, Kiwi, flightless, nocturnal birds endemic to New Zealand, are nearly blind. They rely on a highly developed sense of smell, with nostrils at the end of their long bill. Most birds do not have a good sense of smell. Hearing is generally well developed. For example, even without light, or any visual cues, barn owls can track their prey by sound.

Birds are both endothermic and homeothermic (maintain a constant internal temperature). They also generally maintain a higher body temperature than mammals, with a temperature generally between 40 and 44 °C. About eighty percent of bird species maintain higher temperature than all but 10% of mammal species.

Birds also generally have a faster heart rate and a greater need for oxygen than mammals, and have a higher metabolic rate. Smaller birds generally have a higher metabolism than larger birds, and have relatively higher energy needs. A hummingbird must feed almost constantly during the day, whereas an emperor penguin goes more than two months without feeding while incubating the couple's egg.

Reproduction

Baby bird unable to fly yet

Although most male birds have no external sex organs, the male does have two testes which become hundreds of times larger during the breeding season to produce sperm. The female's ovaries also become larger, although only the left ovary actually functions.

In the males of species without a phallus, sperm is stored within the proctodeum compartment within the cloaca prior to copulation. During copulation, the female moves her tail to the side and the male either mounts the female from behind or moves very close to her. He moves the opening of his cloaca, or vent, close to hers, so that the sperm can enter the female's cloaca, in what is referred to as a cloacal kiss. This can happen very fast, sometimes in less than one second.

A juvenile Laughing Gull.

The sperm is stored in the female's cloaca for anywhere from a week to a year, depending on the species of bird. Then, one by one, eggs will descend from the female's ovaries and become fertilized by the male's sperm, before being subsequently laid by the female. The eggs will then continue their development in the nest.

Many waterfowl and some other birds, such as the ostrich and turkey, do possess a phallus. Except during copulation, it is hidden within the proctodeum compartment within the cloaca, just inside the vent. The avian phallus is purely a copulatory organ and is not used for expelling urine.

After the eggs hatch, parent birds provide varying degrees of care in terms of food and protection. Precocial birds—those that are relatively mature at birth—can care for themselves independently within minutes of hatching. Altricial hatchlings—those born helpless—are blind and naked, and require extended parental care. The chicks of many ground-nesting birds, such as partridges, chickens, and waders, are often able to run virtually immediately after hatching; such birds are referred to as nidifugous. The young of hole-nesters, on the other hand, are often totally incapable of unassisted survival. The process whereby a chick acquires feathers until it can fly is called "fledging."

Mating Systems and Parental Care

The three mating systems that predominate among birds are polyandry, polygyny and monogamy. Monogamy is seen in approximately 91% of all bird species. Polygyny constitutes 2% of all birds, and polyandry is seen in less than 1%.

The jabiru (Jabiru mycteria), the green-and-red macaw, or green-winged macaw (Ara chloroptera), the bald eagle (Haliaeetus leucocephalus), and the albatross (family Diomedeidae) are examples of birds that mate for life.

Monogamous species of males and females pair for the breeding season. In some cases, the individuals may pair for life. For example, the albatross, bald eagle, and Canadian goose, as well as species of macaws, owls, and crows, mate for life. When the mate dies, there is often a re-mating of the surviving bird. There are instances of one bird showing signs of distress upon the death of a mate, although when the mate is removed from the sight of the survivor, this behavior subsides.

One explanation for the high rate of monogamy among birds is the fact that male birds are just as adept at parental care as females. In most groups of animals, male parental care is rare, but in birds it is quite common; in fact, it is more extensive in birds than in any other vertebrate class. In birds, male care can be seen as important or essential to female fitness. "In one form of monogamy, such as with obligate monogamy, a female cannot rear a litter without the aid of a male".

These redwing hatchlings are completely dependent on parental care.

The parental behavior most closely associated with monogamy is male incubation. Interestingly, male incubation is the most confining male parental behavior. It takes time and also may require physiological changes that interfere with continued mating. For example, in the case of the emperor penguin, after the female penguin lays its egg, she returns to the sea to feed for two months to replenish her nutritional reserves. During this time, the male incubates the egg in its brood pouch for about two months consecutively without food. If the chick hatches before the mother's return, the father feeds it a substance produced by a gland in his esophagus.

Monogamy does not necessarily translate to fidelity among the pairs, as examples to the contrary are known. It is certainly problematic to compare mating habits of species in creation with human habits, given that people are endowed with a spiritual as well as physical nature. Nevertheless, the dominance of monogamy in birds, and the extensive phenomena of pairing for life as a universal standard within certain bird species, provides a counterbalance to the current social Darwinistic views of some that monogamy, pairing for life, and fidelity in humans is an unnatural state.

5

Herpetology and Batrachology

The domain of biology which studies the amphibians and reptiles is known as herpetology. Batrachology is a sub-discipline of herpetology which focuses on amphibians such as salamanders and frogs. The topics elaborated in this chapter will help in gaining a better perspective about these animals studied under herpetology and batrachology.

Herpetology

Herpetology is the branch of zoology dealing with the study of reptiles and amphibians.

The study of reptiles and amphibians reveals humanity's strong and lasting curiosity about nature. Since the dawn of humanity, humans have studied reptiles and amphibians, if only as a source of food or because of the dangers of toxins or venoms. Herpetology not only offers many practical benefits—medicinal applications, conservation, environmental indicators, model organisms, and so forth—but also addresses people's desire to better understand the world around them.

Herpetology is concerned with poikilothermic, or cold-blooded, tetrapods (four-legged vertebrates). "Herps" include reptiles like lizards, snakes, crocodiles, and turtles, and amphibians like frogs, toads, salamanders, newts, waterdogs, mudpuppies, sirens, and caecilians. Herpetology is concerned with the study of such extant taxa and even extinct taxa.

Reptiles and amphibians share poikilothermy, and sometimes a superficial resemblance to one another (e.g. salamanders and lizards are often confused), but these two groups are quite distinct. The essential difference is that amphibians include all tetrapods that do not have amniotic eggs, while reptiles are tetrapods and amniotes (animals whose embryos are surrounded by an amniotic membrane that encases it in amniotic fluid). There are many other differences. For example, amphibians tend to have permeable skin that allows for the exchange of gases, are often bound to water for at least part of their lives, have glandular skin, and many of them produce toxic secretions in some of their skin glands, warding off predators and generally tasting bad. Reptiles, by contrast, typically have a dry watertight skin, usually protected by scales, that normally has few if any glands. Many species of reptiles spend no part of their lives anywhere near water, and they all have lungs. Living reptilians usually, if not always, lay eggs or give birth on land, including marine turtles, which only come ashore for this purpose. Again, extinct creatures may have exhibited some differences.

Despite these differences between amphibians and reptiles, they form a common focus under the umbrella of herpetology. Furthermore, despite the differences between these two groups and fish (non-tetrapod vertebrates), it also is not uncommon for herpetological and ichthyological (study of fish) scientific societies to "team up," publishing joint journals and holding joint conferences in order to foster the exchange of ideas between the fields One of the most prestigious herpetology societies is an example of this: the American Society for Ichthyology and Herpetology.

Importance

Because reptiles and amphibians offer important economic, ecological, aesthetic, and medicinal values, herpetology offers numerous benefits to humanity.

For example, amphibians are often very sensitive to environmental changes, offering a visible warning to humans that significant changes are taking place.

Some toxins and venoms produced by reptiles and amphibians are useful in human medicine. Currently, some snake venom has been used to create anti-coagulants that work to treat stroke victims and heart attack cases. Medicine produced from amphibians now are used to treat heart aliments, bacterial infections, skin and colon cancers, depression, and chronic pain, among others. Amphibian toxins continue to be studied by herpetologists for possible applications. For example, a poison frog from South America (Epipedobates tricolor) secretes a non-addictive painkiller 200 times more potent than morphine, offering promise in this area if the toxicity can be neutralized.

The axolotl serves as model organism for research.

Reptiles and amphibians are a critical element in the food chains of most ecosystems, and sometimes a keystone species whose removal can drastically alter the populations of other organisms. The consumption by reptiles and amphibians of rodents and insect pests aid in control of these animals, which can be serious agricultural pests; furthermore, reducing the numbers of insects can lower the prevalence of insect-borne diseases. The skin of crocodilians, snakes, and lizards are used in leather goods, such as shoes, handbags, gloves, and belts, and thus study of these animals can offer direct economic applications.

Some reptiles likewise present threats to people, whether because they are venomous, like some snakes, or can attack humans, such as some crocodilians, or their skin harbors salmonella, a bacterial disease, and thus their study aids in preventing such dangers.

Reptile

Reptile is any member of the class Reptilia, the group of air-breathing vertebrates that have internal fertilization, amniotic development, and epidermal scales covering part or all of their body. The major groups of living reptiles—the turtles (order Testudines), tuatara (order Rhynchocephalia [Sphenodontida]), lizards and snakes (order Squamata), and crocodiles (order Crocodylia, or Crocodilia)—account for over 8,700 species. Birds (class Aves) share a common ancestor with crocodiles in subclass Archosauria and are technically one lineage of reptiles, but they are treated separately.

Common king snake (Lampropeltis getula).

The extinct reptiles included an even more diverse group of animals that ranged from the marine plesiosaurs, pliosaurs, and ichthyosaurs to the giant plant-eating and meat-eating dinosaurs of terrestrial environments. Taxonomically, Reptilia and Synapsida (a group of mammal-like reptiles and their extinct relatives) were sister groups that diverged from a common ancestor during the Middle Pennsylvanian Epoch (approximately 312 million to 307 million years ago). For millions of years representatives of these two groups were superficially similar. However, slowly lifestyles diverged, and from the synapsid line came hairy mammals that possessed an endothermic physiology and mammary glands for feeding their young. All birds and some groups of extinct reptiles, such as selected groups of dinosaurs, also evolved an endothermic physiology. However, the majority of modern reptiles possess an ectothermic physiology. Today only the leatherback sea turtle (Dermochelys coriacea) has a near-endothermic physiology. So far no reptile, living or extinct, has developed specialized skin glands for feeding its young.

General Features

Most reptiles have a continuous external covering of epidermal scales. Reptile scales contain a unique type of keratin called beta keratin; the scales and interscalar skin also contain alpha keratin, which is a trait shared with other vertebrates. Keratin is the main component of reptilian scales. Scales may be very small, as in the microscopic tubercular scales of dwarf geckos (Sphaerodactylus), or relatively large, as in the body scales of many groups of lizards and snakes. The largest scales are the scutes covering the shell of a turtle or the plates of a crocodile.

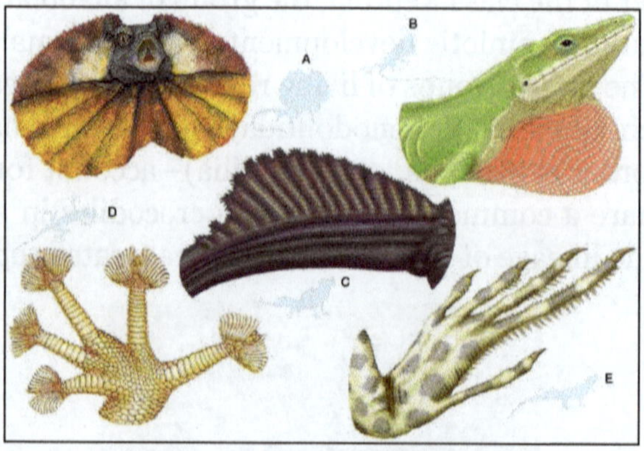

Specialized body structures of lizards: (A) frill of a frilled lizard (Chlamydosaurus kingii), (B) dewlap of an anole (Anolis), (C) fin of the water lizard (Hydrosaurus), (D) toe fans of a fan-footed gecko (Ptyodactylus hasselquistii), and (E) toes of the fringe-toed lizard (Uma).

Other features also define the class Reptilia. The occipital condyle (a protuberance where the skull attaches to the first vertebra) is single. The cervical vertebrae in reptiles have midventral keels, and the intercentrum of the second cervical vertebra fuses to the axis in adults. Taxa with well-developed limbs have two or more sacral vertebrae. The lower jaw of reptiles is made up of several bones but lacks an anterior coronoid bone. In the ear a single auditory bone, the stapes, transmits sound vibrations from the eardrum (tympanum) to the inner ear. Sexual reproduction is internal, and sperm may be deposited by copulation or through the apposition of cloacae. Asexual reproduction by parthenogenesis also occurs in some groups. Development may be internal, with embryos retained in the female's oviducts, and embryos of some species may be attached to the mother by a placenta. However, development in most species is external, with embryos enclosed in shelled eggs. In all cases each embryo is encased in an amnion, a membranous fluid-filled sac.

Importance

In the agriculture industry as a whole, reptiles do not have a great commercial value compared with fowl and hoofed mammals; nonetheless, they have a significant economic value for food and ecological services (such as insect control) at the local level, and they are valued nationally and internationally for food, medicinal products, leather goods, and the pet trade.

Reptiles have their greatest economic impact in some temperate and many tropical areas, although this impact is often overlooked because their contribution is entirely local. A monetary value is often not assigned to any vertebrate that provides pest control. Nonetheless, many lizards control insect pests in homes and gardens; snakes are major predators of rodents, and the importance of rodent control has been demonstrated repeatedly when populations of rodent-eating snakes are decimated by snake harvesting for the leather trade. The absence of such snakes allows rodent populations to explode. Similarly, turtles, crocodiles, snakes, and lizards are regularly harvested as food for local consumption in many tropical areas. When this harvesting becomes commercial, the demands on local reptile populations commonly exceed the ability of species to replace themselves by normal reproductive means. Harvesting is often concentrated on the larger individuals of most species, and these individuals are often the adult females and males in the population; their removal greatly reduces the breeding stock and leads to a precipitous population decline.

The overharvesting of crocodiles for the leather industry in the 1950s and 1960s caused the widespread extirpation, or localized extinction, of many crocodilian species. In addition, surviving populations experienced a near-worldwide drop in numbers. Since then, regulations at the national and international levels have greatly reduced the harvests, and proactive conservation and management measures have allowed many crocodilian populations to rebound. Regulated harvesting currently provides an adequate number of skins to the leather trade and also allows crocodiles to resume their role as top predators in many aquatic ecosystems.

Regulated harvesting of large snakes and lizards is also underway in parts of Indonesia. In addition, several groups of reptiles (tegu lizards in Argentina, freshwater turtles in China, and green iguanas [Iguana iguana] in Central America) are raised as livestock. Often the process of regulated harvesting begins with the removal of a few eggs, juveniles, or adults from wild populations. Stocks of reptiles are raised on farms and ranches. Farms and ranches then sell some individuals to commercial interests, while others are retained as breeding stock.

Size Range

Most reptiles are measured from snout to vent (that is, the tip of the nose to the cloaca). However, measurements of total length are common for larger species, and shell length is used to gauge the size of turtles. The body size of living reptiles varies widely. Dwarf geckos (Sphaerodactylus parthenopion) are the smallest reptiles and have a snout-to-vent length of 16–18 mm (0.6–0.7 inch). In contrast, giant turtles, such as the leatherback sea turtles (Dermochelys coriacea), possess shell lengths of nearly 2 metres (about

7 feet). In terms of total length, the largest living reptiles are the reticulated pythons (Python reticulatus) and saltwater crocodiles (Crocodylus porosus), which may grow to more than 7 metres (23 feet) as adults. Some ancient reptile groups had members that were the largest animals ever to live on the land—some sauropod dinosaur fossils measure 20–30 metres (66–98 feet) in total length. The largest marine reptiles, the pliosaurs, grew to 15 metres (50 feet) long.

The Komodo dragon (Varanus komodoensis) is the world's largest lizard.

The reptile groups also show a diversity of morphologies. Some groups, such as most lizards and all crocodiles, possess strongly developed limbs, whereas other groups, such as the worm lizards and snakes, are limbless. Reptilian body flexibility ranges from the highly flexible forms found in snakes to the inflexible armoured bodies of turtles. In addition, the tails of most turtles tend to be short, especially when compared with the long heavy tails of crocodiles.

Giants in any animal group always attract attention and are often exaggerated. Anacondas (Eunectes), gigantic snakes from South America, are undoubtedly the largest living snakes. The largest species, the green anaconda (E. murinus), likely only rarely exceeds 9 metres (30 feet) in length; nonetheless, persistent but unsubstantiated reports have been made of anacondas that are 12 metres (40 feet) long. The reticulated python (P. reticulatus) of Southeast Asia and the East Indies has been recorded at 10.1 metres (33.3 feet).

At about 5.5 metres (18 feet), the king cobra (Ophiophagus hannah) of Asia and the East Indies is the longest venomous snake. The heaviest venomous snake is probably the eastern diamondback rattlesnake (Crotalus adamanteus); its length does not exceed 2.4 metres (7.9 feet), but it may weigh as much as 15.5 kg (34 pounds). The largest of the common nonvenomous snakes is probably the keeled rat snake (Ptyas carinatus), at about 3.7 metres (12 feet).

Five species of crocodiles may grow larger than 6 metres (20 feet). Nile crocodiles (Crocodylus niloticus) and estuarine (or saltwater) crocodiles (C. porosus) regularly exceed this length. The American crocodile (C. acutus), the Orinoco crocodile (C. intermedius), and the gavial (Gavialis gangeticus) may also grow larger than 6 metres;

however, this is less common. The gavial normally attains a length of about 4–5 metres (12–15 feet).

Eastern diamondback rattlesnake (Crotalus adamanteus).

The giant among living turtles is the marine leatherback sea turtle (D. coriacea), which reaches a total length of about 2.7 metres (9 feet) and a weight of about 680 kg (1,500 pounds). The largest of the land turtles is a Galápagos tortoise (Geochelone nigra), weighing about 255 kg (560 pounds).

The largest modern lizard, the Komodo dragon (Varanus komodoensis) of the East Indies, is a monitor lizard that attains a total length of 3 metres (10 feet). In addition, two or three other species of monitors reach 1.8 metres (5.9 feet). The water monitor (V. salvator) may grow to a greater total length than the Komodo dragon, but it does not exceed the latter in weight. The green iguana, which grows to a length of about 2 metres (7 feet) comes close to that size, but no other lizard does.

Within each reptile group, with the possible exception of snakes, no living member is as large as its largest extinct representative. At about 2.7 metres (9 feet) in total length, the leatherback sea turtle is smaller than Archelon, a genus of extinct marine turtles from the Late Cretaceous (100 million to 65.5 million years ago) that was about 3.6 metres (11.8 feet) long. No modern crocodile approaches the estimated 15-metre (49-foot) length of Phobosuchus, and the Komodo dragon does not compare with Tylosaurs, a mosasaur that exceeded 6 metres (20 feet) in length. Some quadrapedal browsing dinosaurs grew to lengths of 30 metres (100 feet) and weights of 91,000 kg (200,000 pounds) or more.

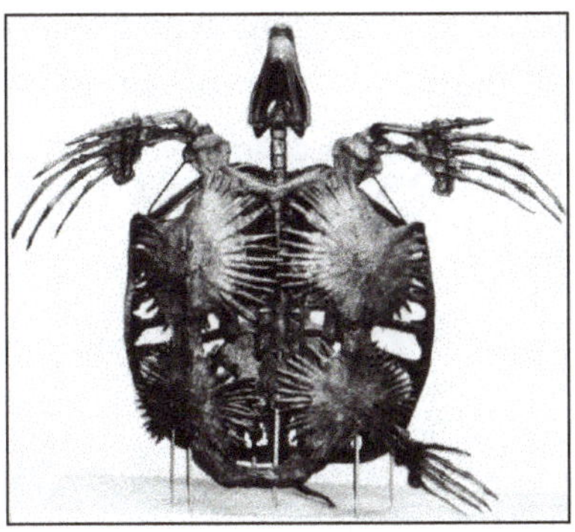

Skeleton of the Cretaceous marine turtle Archelon, length 3.25 metres (10.7 feet).

The smallest reptiles are found among the geckos (family Gekkonidae), skinks (family Scincidae), and microteiids (family Gymnopthalmidae); some of these lizards grow no longer than 4 cm (about 2 inches). Certain blind snakes (family Typhlopidae) are less than 10 cm (4 inches) in total length when fully grown. Several species of turtles weigh less than 450 grams (1 pound) and reach a maximum shell length of 12.5 cm (5 inches). The smallest crocodiles are the dwarf crocodiles (Osteolaemus tetraspis), which grow to about 2 metres (7 feet), and the dwarf caimans (Paleosuchus), which typically grow to 1.7 metres (6 feet) or less.

Day gecko (genus Phelsuma).

Potential predators of the blue-tailed skink (Plestiodon skiltonianus) are attracted to its tail, which can be shed at will.

Distribution and Ecology

Reptiles are mainly animals of Earth's temperate and tropical regions, and the greatest number of reptilian species live between 30° N and 30° S latitude. Nevertheless, at least two species, the European viper (Vipera berus) and the common, or viviparous, lizard (Lacerta vivipara, also called Zootoca vivipara), have populations that edge over the Arctic Circle (66°33′39″ N latitude). Other species of snakes, lizards, and turtles also live at high latitudes and altitudes and have evolved lifestyles that allow them to survive and reproduce with little more than three months of activity each year.

European viper, or common adder (Vipera berus).

Reptile activity is strongly dependent on the temperature of the surrounding environment. Reptiles are ectothermic—that is, they require an external heat source to elevate their body temperature. They are also considered cold-blooded animals, although this label can be misleading, as the blood of many desert reptiles is often relatively warm. The body temperatures of many species approximate the surrounding air or the temperature of the substrate, hence a reptile can feel cold to the human touch. Many species, particularly lizards, have preferred body temperatures above 28 °C (82 °F) and only pursue their daily activities when they have elevated their body temperatures to those levels. These species maintain elevated body temperatures at a relatively constant level by shuttling in and out of sunlight.

Reptiles occur in most habitats, from the open sea to the middle elevations in mountainous habitats. The yellow-bellied sea snake (Pelamis platurus) spends all its life in marine environments. It feeds and gives birth far from any coastline and is helpless if washed ashore, whereas other sea snakes live in coastal waters of estuaries and coral reefs. The sea turtles are also predominately coastal animals, although most species have a pelagic, or open-ocean, phase that lasts from the hatchling stage to the young juvenile stage. Many snakes, crocodiles, and a few lizards are aquatic and live in freshwater habitats ranging from large rivers and lakes to small mountain streams. On land, turtles, snakes, and lizards also occur widely in forests, in grasslands, and even in true deserts. In many arid lands lizards and snakes are the major small-animal carnivores.

North Temperate Zone

Reptiles of the north temperate zone include many ecological types. Aquatic groups are represented in both hemispheres by the water snakes, many testudinoid turtles, and the two species of Alligator. Terrestrial groups include tortoises, ground-dwelling snakes, and many genera of lizards. Arboreal snakes are few, and arboreal lizards are almost nonexistent. There are few specialized burrowing lizards in this region, but burrowing snakes are common.

The viviparous lizard (L. vivipara, or Z. vivipara) and the European viper (V. berus) are the most northerly distributed reptiles. A portion of each reptile's geographic range occurs just north of the Arctic Circle, at least in Scandinavia. Other reptiles—the slowworm (Anguis fragilis), the sand lizard (L. agilis), the grass snake (Natrix natrix), and the smooth snake (Coronella austriaca)—also appear at high latitudes and reach to 60° N in Europe. Of these six northern species, all but the grass snake are viviparous (live-bearing). Across Siberia only L. vivipara and V. berus live north of 60° N.

In North America no reptile is found at 60° N latitude or higher. Two species of garter snakes (Thamnophis) live as far north as 55° N in western Canada. However, it is only south of 40° N that numerous species of reptiles occur. In the eastern United States and eastern Asia, several colubrid snake species, northern skinks (Plestiodon), glass lizards (Ophisaurus), and softshell turtles (Trionychidae) are common.

Across North America and Eurasia, the northern limit of turtles is about 55° N. Even though these regions are characterized by many species of turtles, most families and genera are unique to one region or another. This phenomenon also occurs in other groups of reptiles. Many lizards of temperate Eurasia belong to the families Agamidae and Lacertidae, which do not occur in the Americas, whereas many lizards of North America are in the families Iguanidae and Teiidae, which do not live in Eurasia. Nonetheless, of the two living species of Alligator, one (A. mississippiensis) lives in the southeastern United States, and the other (A. sinensis) lives in China.

The reptiles of the eastern United States are almost as distinct from those of the western United States and northern Mexico as they are from those of eastern Asia. The eastern region (that is, the eastern United States) has many genera and species of emydid turtles. In contrast, the western region (that is, the western United States and northern Mexico), which is defined by a diagonal line running southeast to northwest through Texas, then northward along the Continental Divide, has only four or five species of emydids. Few genera and species of iguanid lizards inhabit the eastern region, whereas the western region has many. Although the eastern region has more species of water snakes, the western region contains more garter snakes. Whereas more species of snakes appear in the eastern United States than in the western areas, the converse is true of lizard species.

Gila monster (Heloderma suspectum).

Central and South America

Nearer to the Equator, reptiles become more numerous and diverse. This is true of crocodiles; Mexico is home to three species, but nine countries in South America are home to four or more crocodilian species. Turtles, lizards, and snakes are also particularly diverse in this region.

Some groups of North American turtles are represented in the New World tropics. For example, the mud turtles (Kinosternon) and sliders (Trachemys) appear in both regions, but the majority of species are members of genera and even families (such as the side-necked turtles of families Podocnemididae and Chelidae) that are unknown in

North America. In addition, Central America has three endemic genera of turtles (Dermatemys, Claudius, and Staurotypus).

Many of the genera of iguanid lizards occurring in the western United States have species in Mexico; one genus of spiny lizards (Sceloporus) is most diverse in Mexico. South of Mexico the North American iguanids disappear and are replaced by tropical groups such as the black iguanas (Ctenosaura), the helmeted iguanids (Corythophanes), the casque-headed iguanids (Laemanctus), and the basilisks (Basiliscus). Iguanid lizards of the anole genus (Anolis) are represented in northern South America and the West Indies by more than 200 species. Other iguanid genera—the long-legged Polychrus—make their appearance.

The lizard family Teiidae, though represented in the United States by the racerunners and whiptails of the genus Cnemidophorus, is primarily tropical, and its area of greatest biological diversity begins in Central America with the large, conspicuous, and active ameivas (Ameiva). The Gymnophthalmidae (or microteiids), close relatives of the teiids, are a diverse group of small-bodied lizards that live largely in and under leaf litter in the West Indies, Central America, and South America north of central Argentina.

Among snakes, fer-de-lances (Bothrops), coral snakes (Micrurus), rear-fanged snakes (such as the cat-eyed snakes [Leptodeira]), and certain nonvenomous genera (such as the tropical green snakes [Leptophis]) do not occur north of Central America. Farther south these groups become more diverse. Vine snakes (Oxybelis and Imantodes), false coral snakes (Erythrolamprus), coral snakes (Micrurus), slender ground snakes (Drymobius), and the burrowing spindle snakes (Atractus) are some of the most biologically diverse groups in this region.

Several groups of reptiles that form important, if not dominant, elements of the fauna of the Eastern Hemisphere are largely or completely absent from the American tropics. Such groups include the lizard families Agamidae, Chamaeleonidae, Lacertidae, and Scincidae and many genera, subfamilies, and families of snakes.

South of the tropics, in the temperate regions of South America, the diversity of reptiles diminishes rapidly. Crocodiles and turtles do not occur south of northern Argentina. However, the range of one viviparous pit viper reaches to almost 50° S, and the ranges of two iguanid lizards—Magellan's tree iguana (Liolaemus magellanicus) and the Cascabel rattlesnake (Crotalus durissus)—extend to almost to 55° S.

Asia

The temperate zone of Eurasia is noted for its many lizards of the families Agamidae and Lacertidae, and, to lesser degrees, Gekkonidae and Scincidae. Most of the lizards are terrestrial, and extremely specialized burrowers include desert-dwelling skinks (Ophiomorphus and Scincus). Most of the snakes characteristic of this vast area are also terrestrial, and the leaf-nosed snakes (Lytorhynchus) and the sand boas (Eryx) are the distinctive burrowing snakes of the region. Arboreal snakes are represented almost exclusively by the rat snakes (Elaphe).

A few types of reptiles characteristic of the Asian tropics extend into the temperate zone—such as several rear-fanged snakes (Boiga trigonata and Psammodynastes), some cobras of the genus Naja, several species of softshell turtles (Trionychidae), and some species of skinks and geckos. Except for the Chinese alligator (Alligator sinensis) and the Indian gavial (Gavialis gangeticus), crocodiles are absent from temperate Eurasia.

In the Asian tropics, the reptilian fauna is extremely rich in species and encompasses several diverse types. Aquatic reptile groups are represented by snakes of various genera (such as Natrix, Enhydris, and Acrochordus), several groups of lizards (Tropidophorus among the skinks and Hydrosaurus among the agamids), many batagurids and soft-shelled turtles, and five species of crocodiles. Asia's numerous terrestrial reptiles include the small kukri snakes (Oligodon), the big Asian rat snakes (Ptyas), cobras (family Elapidae), monitor lizards (Varanus), many species and genera of skinks, some geckos, and several land turtles (Cuora, Indotestudo, and Geochelone). Specialized burrowing snakes (such as those of family Uropeltidae and the colubrid genus Calamaria) and lizards (such as the family Dibamidae and the skink genus Brachymeles) contain many.

Many distinctive life forms of reptiles in tropical Asia are arboreal. They include pythons and Asian pit vipers (Trimeresurus), vine snakes (Ahaetulla), slug-eating snakes (Pareas), flying snakes (Chrysopelea), and tree racers (Gonyosoma). Some lizards, such as the monitors, climb only with the aid of claws. A few others—such as the deaf agamids, Cophotis—climb with the help of prehensile, or grasping, tails. Other Asian reptiles, such as several species of geckos, climb with the help of clinging pads under the digits. The most striking arboreal reptiles of this area are the flying lizards (Draco), which possess spreadable rib wings, and the parachuting gecko (Ptychozoon), which has fully webbed digits, a fringed tail, and wide flaps of skin along its sides.

Australia

As a result of close geological relationships and faunal similarities, the general characteristics of reptiles in the Australian faunal region also apply in New Guinea. Australia is the only continent in the world in which venomous snake species outnumber nonvenomous ones. The family Colubridae, which encompasses the majority of the nonvenomous or slightly venomous snakes of the world, is poorly represented in Australia, with fewer than a dozen species. The reptilian fauna also includes several pythons and minute blind snakes (family Typhlopidae); a great variety of geckos, skinks, and agamid lizards; side-necked turtles (family Chelidae); and three species of crocodiles. The Australian region is home to a diverse group of cobras (family Elapidae) but no vipers.

Africa

The reptilian fauna of Africa has two main components. The first, the fauna of North Africa, is akin to that of central and southwestern Asia and southern Europe and thus is mainly a Eurasian fauna. The racers (Coluber), the burrowing sand skink (Scincus), and the batagurid turtle (Mauremys caspica) are elements of this fauna in North Africa.

North African reptiles, though representing many families, are principally terrestrial and burrowing. Many lacertid and agamid lizards scamper over rocks and sand by day; they are replaced at night by small geckos and are preyed upon by the racers and sand snakes. In addition to cobras, the venomous snakes of North Africa include the common vipers, the saw-scaled viper (Echis carinatus), and the horned vipers (Cerastes). The last two are true desert animals and also occur in Southwest Asia.

Some reptilian genera from sub-Saharan Africa also occur in North Africa and in southwestern Asia. Examples include the sand snakes (Psammophis), cobras, and chameleons (family Chamaeleonidae).

The second and much larger component of the African herpetofauna is the sub-Saharan assemblage that ranges from the Sahara southward to the Cape of Good Hope. In common with tropical Asia, this vast area is home to cobras and many skinks and geckos. Its fauna differs from that of Asia by the absence of pit vipers (subfamily Crotalinae), the absence of batagurid turtles, and the presence of a few agamid lizards. These groups are replaced in tropical Africa by the many true vipers (subfamily Viperinae), the side-necked turtles (family Pelomedusidae), the wall lizards (family Lacertidae), and the spiny-tailed lizards (family Cordylidae). Numerous species of chameleons and tortoises and three species of crocodiles occur in sub-Saharan Africa.

The large island of Madagascar, off the eastern coast of Africa, has a peculiar fauna with its affinities mainly to African reptile groups. Because of its long isolation, Madagascar possesses distinct genera and subfamilies of chameleons and other reptiles.

Life Cycle and Life History

The diversity of reptile life histories is amazingly wide and often reveals nearly unimaginable reproductive adaptations. Some reptiles are annual species that hatch, mature, reproduce, and die in one year or, at most, two years (as in side-blotched lizards [Uta stansburiana]). Others, such as loggerhead sea turtles (Caretta caretta), are long-lived species that require 25 or more years to mature and have life spans that exceed 50 years. Numerous other species fall between these extremes. Some reptiles lay eggs, whereas others are live-bearers. Some species lay 1 or 2 eggs, whereas others lay 100 or more eggs in each nesting event. Some reptiles nest year round, whereas others may nest once each year or allow two or more years between breeding cycles.

Courtship and Fertilization

The evolution of amniotic development and the shelled egg enabled vertebrates to become fully terrestrial. These two evolutionary advances required the previous development of internal fertilization. In other words, the deposition of sperm by the male into the female's reproductive tract and the sperm's subsequent penetration of the egg cell was necessary before the shelled egg could exist.

In living reptiles the deposition of the male's sperm inside the body of the female occurs by cloacal apposition or the use of an intromittent, or copulatory, organ. The former method is characteristic of only one group, the tuatara (Sphenodon), which copulate via the close alignment of the male's cloaca (that is, a common chamber and outlet into which the intestinal, urinary, and genital tracts open) with that of the female. The male then discharges semen into the female's cloaca. In all other reptiles, males have either a penis (as in turtles [order Testudines] and crocodiles [order Crocodylia, or Crocodilia]) or hemipenes (as in lizards and snakes [order Squamata]). The penis is a homologue of the mammalian penis, and its presence in reptiles indicates that this organ arose early in the evolution of the amniotes and prior to the origin of reptiles and synapsids. In contrast, the hemipenes are structurally quite different. They are labeled "hemi" because two occur in each male, although only one is used during a single copulatory event. Whether a penis or hemipenis, this organ is inserted into the female cloaca.

Once semen is deposited, the sperm must move out of the female's cloaca and into each oviduct. They move up the oviduct to an opening adjacent to an ovary. The mechanism of how the sperm find this pathway remains largely unknown, but for successful fertilization the sperm must be above the oviduct glands that will secrete the shell of the egg. When ovulation occurs, the eggs are shed from the ovary and drop directly into the oviduct, one on each side. In reptiles copulation may stimulate ovulation, occur simultaneously with ovulation, take place within an hour to a week of ovulation (presumably the most frequent situation), or occur months prior to complete the development of the eggs and their ovulation.

Although spring is the main period of courtship and copulation for most temperate-zone reptiles, males commonly complete spermatogenesis (that is, the production of sperm) in late summer. Occasionally, a male will mate, and his sperm will be stored in the oviducts of the female until the eggs are ovulated in the spring (such as in snapping turtles [Chelydra serpentine]). This ability to store sperm seems to be widespread in snakes and turtles, although the phenomenon has not been rigorously tested. One study showed that the diamondback terrapin (Malaclemys terrapin) could produce viable eggs four years after copulation, although the percent of fertile eggs declined sharply after one year and progressively to the fourth and final year of the experiment.

For a successful copulation to occur, cooperation between the female and male is required. In most reptiles the male courts the female with a series of behaviours to assess her reproductive readiness and receptivity. Many lizards also have a distinct pattern of head bobs and forebody push-ups. Combined with water vibrations and sprays, male crocodiles also use body movements to court females and warn off other males. In anoles (Anolis) and flying lizards (Draco), males have well-developed and brightly coloured throat fans, or dewlaps, that open and close. Throat fans are used to attract females and play a large role in territorial disputes with other males. Turtles use visual and olfactory displays and tactile cues in courtship. These signals occur in various combinations and are species-dependent. For example, in some turtle species the female seems to be pestered into submission.

Courtship in snakes and many scleroglossan lizards may also involve the use of pheromones that ensure that courtship and copulation occur between members of the same species. Pheromones may also help to attract a member of the opposite sex and thus illicit the female's cooperation in the reproductive process. Snakes rely mainly on pheromone and tactile stimulation. The male crawls over the female and regularly taps his chin on her back; this behaviour presumably results in an exchange of pheromones, which simultaneously stimulates the participants.

The so-called courtship dance of many snakes is often mistakenly interpreted as a dance in which the forepart of the bodies of a male and a female are held high and entwined. It is actually a power struggle between two males competing over the same female. The goal of the courtship dance is to press the body of the opponent to the ground. The swaying wrestling match continues until one male concedes defeat and crawls away. Often by that time the female, who was probably in the midst of being courted by one of the males, has departed, and she must be tracked (through her odour trail) by the victorious male for copulation to occur.

Other male reptiles also decide dominance and access to females by combat. Monitor lizards (Varanus) wrestle, truly grappling with one another; they may stand on their hind limbs and tail to attempt to force one another to the ground. Among turtles, male tortoises commonly ram each other with their heads and bodies. The objective is to drive away the opposing male, and it is best if an opponent can be rolled onto his back. A pair of mated sea turtles is regularly accompanied by other males that bite at the mounted male in an effort to displace him. The male's objective is to ensure that his sperm fertilizes the female's egg so that the offspring will share his genes. One method, aside from combat, to ensure limited insemination of the female is through the deposition of a mucous copulatory plug. Male garter snakes (Thamnophis) deposit this plug into the female's cloaca at the end of copulation. The plug prevents any other mating and remains for a day or two.

In a few species of lizards—including certain geckos (Gekkonidae), racerunners (Tediidae), rock lizards (Lacertidae), monitor lizards (Varanus), and the brahminy blind snake (Ramphotyphlops braminus)—females may reproduce by parthenogenesis (that is, their eggs require no sperm activation or fertilization). Instead, the eggs are self-activated and spontaneously begin cell division and differentiation once they are ovulated and deposited in a nest. In many cases the entire species is unisexual and contains only females. In Komodo dragons (V. komodoensis) and other bisexual species, some females may reproduce parthenogenetically, whereas other females reproduce sexually. Because no sperm are used, male chromosomes are not available, and recombination does not occur. Consequently, the resulting offspring have the same genetic makeup as the mother. In unisexual species such as R. braminus all individuals have the same genetic composition, and the entire species is likely to have arisen from one female.

Embryonic Development and Parental Care

Once the eggs are fertilized, development begins, and the egg becomes an embryo as it divides into successively smaller cells. The time delay between fertilization and egg deposition (that is, egg laying) is poorly documented for the majority of reptile species. Whereas copulation and the delivery of sperm into the female's reproductive tract can occur weeks or months before the eggs are ovulated, fertilization and egg deposition typically appear to occur within hours to days of ovulation. Apparently, many egg-laying (or oviparous) reptiles have a mechanism to retard or stop development in the oviduct once the early gastrula stage is attained. However, in most species, development continues as soon as the egg is deposited. During periods of high stress and other relatively unusual conditions (such as in captivity), females have been known to retain shelled eggs in their oviduct for weeks to months. In some situations where protracted egg retention results, eggs have eroded the oviductal wall and have fallen into the body cavity.

Egg-laying and nest-building behaviours vary widely among reptiles. These behaviours range from the "casual" dropping of the eggs in a relatively suitable site to the preparation of an elaborate nest, and in a few groups parental care may also occur. Most turtles dig an egg chamber exclusively with their hind limbs, and attention is given to the selection of the nest site, the excavation of the egg chamber, and its closure. Thereafter, the female departs, and the eggs and hatchlings must survive on their own. Most lizards and snakes also depart after the eggs are lain; the egg chamber can be little more than a hollow as the lizard or snake crawls through leaf litter or soil, or it may be more elaborate. For example, the common, or green, iguana (I. iguana) digs a deep burrow with a combination of its fore- and hind limbs; this chamber is often so deep that the female is totally hidden from view. At the end of this burrow, she lays her eggs and fills the entire burrow with loosened soil. Often a group of females will return to the same nesting site within the same nesting colony year after year.

Snakes can also dig elaborate and deep chambers; the pine snake (Pituophis melanoleucus) lives on sandy soil and uses its head and the forepart of its body to scoop soil from its burrows and egg chambers. Many geckos deposit their eggs in cracks or crevices in rock faces, in tree bark, or in plant tissues beneath the bark of trees. The eggs of some geckos are adhesive and may be attached to vertical surfaces; in other geckos several females will share a good nesting site beneath a slab of rock or behind the loose bark on the side of a tree. Such locations may contain dozens of eggs at different stages of development.

Although a few species of lizards and snakes remain with their clutch—often curling their bodies around their eggs for the entire duration of the incubation period—the most intricate examples of parental care occur in crocodiles. Even though there are species-specific variants in behaviour, the female crocodile typically creates a nest mound of soil and vegetation, using her mouth, limbs, body, and tail in its construction. After she digs a hole in the mound and lays her eggs, her attention remains focused on her eggs, and she stays nearby to watch over them. As the eggs begin to hatch, the hatchlings begin to chirp and squeak, bringing their mother to the nest. She uncovers the eggs and may even use her tongue to help some of the hatchlings out of their eggshells. She then carries her young to the water in her mouth and will stay with them for several months until they are large enough to survive on their own.

Some reptiles may bear their young alive. This mode, called viviparity, is widespread and has evolved independently dozens of times in the squamates (that is, the lizards and snakes). No living crocodiles, turtles, or tuatara are live-bearers. However, in the squamates, live-bearing ranges from retention of unshelled eggs in the oviducts to the development of placentae between the mother and her fetuses. The evolutionary steps from egg laying to placental development are demonstrated by extant species. For example, the rough green snake (Opheodrys aestivus) retains eggs for varying periods, and it can deposit eggs containing full-term embryos that hatch within days of deposition. In other taxa the eggs are not shelled but remain in the oviducts throughout development. The yolk nourishes each embryo, although gas exchange does occur across the amniotic membranes and the oviductal walls. Placental development ranges from simple wall contact and gas exchange between the mother and a developing embryo to the full interdigitation of maternal and fetal tissue for nutrition and gas exchange (as in garter snakes [Thamnophis]). There are several types of placentae that have evolved in squamates that use various components of the amniotic membranes.

Clutches of eggs and litters of neonates vary widely in reptiles and are species-dependent. Among egg layers a clutch may range from a single egg to more than 100. Among live-bearing reptiles, a litter may range from 1 to about 50 neonates. Adult body size is just one aspect associated with number of offspring; genetic constitution and nutrition are also major factors.

The smallest of the living reptiles typically have the fewest offspring, often laying only one or two eggs or producing only one or two neonates. Many geckos and some skinks have genetically fixed clutch sizes of two eggs, and one egg is usually produced by each ovary during a given reproductive cycle. Conversely, turtles and crocodiles produce some of the largest clutches among living reptiles; sea turtles often produce more than 100 eggs each time, whereas the larger crocodiles average 40–50 eggs per clutch. Some of the larger snakes also produce clutches or litters of 40–50 eggs or embryos, but most squamates, even large-bodied species, produce less than 20 eggs or embryos during each reproductive cycle.

Nutrition clearly affects the number of offspring produced, with malnourished females laying fewer eggs or giving birth to fewer offspring. A female lizard suffering through a drought year or coping with loss of her tail may resorb maturing egg follicles in the ovary or forego egg development altogether during that year.

The frequency of reproduction is also governed by energy availability. Female timber rattlesnakes (Crotalus horridus) commonly breed every third year because the female eats little during the summer of her pregnancy. She requires the following summer to rebuild her fat (energy) stores for the subsequent year's pregnancy and egg development.

The duration of egg incubation and pregnancy is temperature-dependent. Because reptiles are ectothermic, the embryos of live-bearing females and the eggs of oviparous females deposited in the soil or other locations are subject to fluctuating temperatures. In general, cool temperatures slow development and warm temperatures speed development, but extreme heat and cold are lethal to developing embryos. On average, temperate-zone reptiles have incubations or pregnancies of 8–12 weeks. Tropical species tend to have similar incubation periods; however, incubations of some species may last nearly one year or longer (as in the Fijian iguana [Brachylophus fasciatus]).

In addition to hereditary, or genetic, factors, the sex of many reptile species may be manipulated by the environment in which embryonic development takes place. Environment-dependent sex determination (ESD) is the collective term for all factors (such as temperature, moisture, and others) that affect the ratio of males to females produced in a given clutch of eggs or a litter of neonates. Temperature-dependent sex determination (TSD), discovered in the early 1970s, is the most researched of these factors. The sex of the offspring in species with TSD is influenced by the temperature during one critical period of incubation, instead of by hereditary factors. In most turtles females are produced at high temperatures and males at low temperatures. At a narrow range of intermediate temperatures, roughly equal numbers of males and females are produced. The reverse occurs in many crocodiles, and females result from cooler temperatures. Some squamates also display TSD, but the sex of most species appears to be primarily determined by genetics.

In egg-laying reptiles the hatchling must break through the eggshell. For this purpose turtles, crocodiles, and tuatara bear a horny pointed caruncle on their snout. The hatchling uses the caruncle to slice open the amniotic membranes and then the eggshell. Squamates have an egg tooth, a special premaxillary tooth that extends forward and out of the mouth, to cut through membranes and shell. Generally, the hatchling rests briefly once out of the shell. If the nest is buried under soil or other material, a hatchling must dig upward to emerge on the surface. Sometimes this occurs in concert with other hatchlings in the nest; a coordinated behaviour is necessary for sea turtles and other species whose eggs are buried deep. In a few species of turtles, such as the North American painted turtle (Chrysemys picta), the hatchlings leave the eggshell, but they remain in the nest through the winter and emerge in the spring. Individual painted turtle hatchlings can tolerate short periods of extreme cold that freezes much of the water in their bodies.

Live-bearing reptiles give birth in the same manner as mammals. If the amniotic membranes do not rupture during birth, the neonate must struggle to break free from the encapsulating membranes.

Growth and Longevity

Reptiles, especially turtles, are noted for their extreme longevity. Many turtles have long lives, but few species have individuals that live more than a century. Records of longevity are derived from captive animals that led protected and catered life. Many North American turtle species require 12 to 18 years to reach sexual maturity. Once they reach adulthood, mortality rates decline substantially, and many individuals reach and exceed 30 years (as in Blanding's turtle [Emydoidea blandingii] and the eastern box turtle [Terrapene carolina]). Generally, the larger the animal, the greater is its life span, so crocodiles, large snakes (such as boas and pythons), and large lizards often live more than 20 years.

Although patterns of growth are poorly documented for the majority of reptiles, most species probably follow a pattern of determinate, or asymptotic, growth as they mature. Most reptiles are characterized by a period of rapid juvenile growth that slows upon reaching full adulthood. Growth then ceases altogether a few years after maturity.

In contrast, some large-bodied species likely have what is known as indeterminate, or attenuated, growth. Typically, rapid growth occurs in juveniles and slows as the individual approaches maturity and shifts its energy resources to reproduction. During most of the adult years, growth is either extremely slow or nonexistent. However, when food resources are high, active growth can occur. Thus, the size of an individual of a species characterized by attenuated growth is only limited by its food supply.

Behaviour

Avoidance and Noise

Avoidance, the most common form of defense in the animal kingdom, is also the most common form of defense in reptiles. At the first recognition of danger, most snakes and lizards crawl or scamper away into the undergrowth; turtles and crocodiles plunge into water and sink out of sight. Even so, should danger arise so suddenly and so close at hand that flight may be hazardous, other behaviours are adopted.

Crocodiles, turtles, some lizards, and some snakes hiss loudly when confronted by an enemy. Rattlesnakes rapidly vibrate the tip of the tail, which consists of loose, dry, horny rings. Even snakes without rattles, such as the fox snake (Elaphe vulpina) of the United States, often rapidly vibrate the ends of their tails. Often, the tail will come into contact with dry leaves, and the resulting sound will seem deceptively like the rattle of a rattlesnake.

Body Form and Posturing

Change in body form is relatively common in snakes. It usually involves spreading the neck, as in the cobras (family Elapidae), or the whole body, as in the harmless hognose snakes (Heterodon) and DeKay's snake (Storeria dekayi) of the United States. Some snakes inflate the forward parts of their bodies; inflation is one of the defensive behaviours of the large South American bird snake Pseustes poecilonotus and the African boomslang (Dispholidus typus).

Black-necked cobra (Naja nigricollis)

Snakes may also assume threat postures as they change their body form. A cobra raises the forepart of its body and spreads its hood when threatened. The typical defensive posture of vipers is the body coiled and the neck held in an S-curve, the head poised to strike.

Some lizards flatten their bodies, puff out their throats, and turn broadside to the enemy. The helmeted iguanids (Corythophanes) of Central America and the chameleons of Africa increase their apparent size in this way when approached by snakes. The Australian bearded lizard (Pogona barbata) spreads its throat downward and outward. The Australian frilled lizard (Chlamydosaurus kingii) suddenly raises a wide membrane, or frill, which extends backward from the throat. Many lizards and snakes open their mouths when threatened but do not strike. A common African lizard, the black-necked agama (Acanthocercus atricollis), faces an enemy with head held high and mouth open to show the brilliant orange interior.

Display of Colour

The display of bright colours is often defensive. This behaviour occurs in some red- or yellow-bellied snakes that turn over or curl up their tails, exposing the brightly coloured undersurface. This behaviour is known in harmless snakes, such as the American ring-necked snake (Diadophis), as well as venomous snakes, such as the Southern coral snake (Micrurus frontalis), with red, orange, or yellow undersides. Although not yet fully understood, these colours must have some significance to predators. Many other animals coloured red, orange, or yellow are either distasteful to predators or possess defenses capable of killing or injuring them. Hence, these colours are thought to serve as warning coloration to potential predators.

Müllerian mimicry: coral snakes

Müllerian mimicry involving (left) the venomous Eastern coral snake Micrurus fulvius, (right) the harmless king snake Lampropeltis polyzone, and (bottom) the moderately venomous rear-fanged false coral snake (Oxyrhopus).

Camouflage that involves both form and colour is common in reptiles. For example, many arboreal snakes and lizards are green; some of the green-coloured snakes, such as the vine snakes of South America (Oxybelis) and southern Asia (Ahaetulla), are very slender and resemble plants common in the habitat. Likewise, lizards of semiarid and rocky habitats are frequently pale and have blotched patterns that resemble pebbles and gravel—as in the leopard lizard (Crotaphytus wislizeni) of the southwestern United States.

Mimicry of dangerous species by harmless ones is a passive defense; however, its validity as an actual mechanism of defense is sometimes challenged. Nonetheless, evidence of mimicry appears among different groups of snakes. For example, the venomous American coral snakes (Micrurus) have various ringed patterns of red, yellow, white, and black. These patterns are matched often by non- or mildly venomous snake species occurring in the same area.

Striking and Biting

If a threatening posture does not succeed in driving off an enemy, many reptiles may become more aggressive. Some snakes (such as DeKay's snake) strike, but with their mouths closed. Others (such as the hognose snakes) strike with their mouth open but do not bite, but snakes of many species will strike and bite viciously. Among the nonvenomous snakes of North America, few are as quick to bite as the water snakes of genus Nerodia; however, they are nonvenomous.

Most of the dangerously venomous snakes (vipers, pit vipers, and cobras) bite in self-defense. Vipers and pit vipers usually strike from a horizontally coiled posture. From this position, the head can be rapidly shot forward, stab the enemy, and be pulled back in readiness for the next strike. From the typical raised posture, a cobra sweeps its head forward and downward to bite. To strike again, it raises its head and neck once more; such aggressive, defensive movements of cobras are slower than those of pit vipers.

Many lizards, regardless of family and size, also bite in defense. For example, the tokay gecko (Gekko gecko) of Southeast Asia bites if sufficiently threatened. Although small lizards have a bite that is effective against only the smallest predators, a large monitor lizard (Varanus) can inflict a painful wound with its large teeth and strong jaws. Some turtles, particularly the softshell turtles of family Trionychidae and snapping turtles of family Chelydridae, bite frequently and vigorously.

Spitting

The spitting of venom by some Asian and African cobras (Naja) and the ringhals (Hemachatus haemachatus) is a purely defensive act directed against large animals. Instead of a straight canal ending in a long opening near the tip of each fang as in most cobras, the specialized fang of the spitting cobra has a canal that turns sharply forward to a small round opening on the front surface. At the moment of ejection, the mouth is opened slightly, and a fine stream of venom is forced out of the fangs by the contraction of the muscle enveloping the poison gland. A spitting cobra usually raises its head and the forepart of its body in the characteristic cobra defensive posture prior to spitting, but venom can be ejected from any position. The effect on skin is negligible; the eyes, however, may be severely damaged, and blindness can result unless the venom is washed out quickly.

Use of the Tail

A few lizards, representing different families, have thick tails covered by large, hard, spiny scales. Such a tail swung vigorously from side to side is an effective defense against snakes, especially when the head and body of the lizard are in a burrow or wedged between rocks.

The tails of some lizard species are useful in defense in another way. When captured, some lizards voluntarily shed, or autotomize, their tails, which wriggle violently, temporarily confusing the predator and allowing the lizard to escape. Each vertebra of the tails of tail-shedding lizards has a fracture plane that can voluntarily split by the appropriate twitch of the tail muscles. Simultaneous stimulation of the nerves in the severed portion keeps it twitching for a few seconds after separation. Usually the tail is broken in only one place, but a few lizards, particularly the so-called glass snakes (Ophisaurus), break their tails into several pieces. The stump heals quickly, and a new tail grows; often, however, the regenerated tail is not as long as the original and has simpler scales.

Snakes, turtles, and crocodiles may have their tails bitten off by predators. However, they cannot break them voluntarily or regenerate them. In confrontations with enemies, some snakes use their tails as diversions by raising them and moving them slowly. Species with this habit commonly have thick, blunt, brightly coloured tails. For example, the small African burrowing python (Calabaria reinhardtii) waves its tail in the air as it moves slowly away from a threat.

Balling

Many snakes, both harmless and venomous, attempt to hide their heads under coils of their bodies. For most species with this habit, the body may be coiled loosely. However, it may also be tightly coiled so that it forms a compact ball with the head in the centre. Balling, as the latter habit is called, is a characteristic response of Calabaria and another African python, Python regius. The African armadillo lizard (Cordylus cataphractus), a species with heavy scales on its head and hard spiny scales covering its body and tail, rolls on its back and grasps its tail in its mouth to present an imposing ring of hard spines to a predator.

Odours

Some reptiles use musk-secreting glands when other defensive measures fail. The water snakes (Nerodia), the garter snakes (Thamnophis), and the alligator lizards (Gerrhonotus) emit a foul-smelling substance from their cloacal glands. An assortment of turtles, such as the mud turtle and the musk turtle (Kinosternidae), have glands on the bridge of their shells that excrete a vile-smelling fluid that likely makes them distasteful to many predators.

Feeding Habits

With few exceptions, modern reptiles feed on some form of animal life (such as insects, mollusks, birds, frogs, mammals, fishes, or even other reptiles). Land tortoises are vegetarians, eating leaves, grass, and even cactus in some cases. The green iguana (I. iguana) of Central and South America, the chuckwalla (Sauromalus obesus) of the southwestern United States and northern Mexico, and the spiny-tailed agamids (Uromastyx) of North Africa and southwestern Asia also are herbivorous. The marine iguana (Amblyrhynchus cristatus) of the Galápagos Islands dives into the sea for seaweed.

The majority of carnivorous reptiles have nonspecialized diets and feed on a variety of animals. In general, the smaller the reptile, the smaller is its prey. Only the very largest of living snakes—the reticulated python (Python reticulatus), the Indian python (P. molurus), and the green anaconda (Eunectes murinus)—are capable of eating large mammals such as small pigs and deer. Among crocodiles the largest species—the Nile crocodile (Crocodylus niloticus), the estuarine, or saltwater, crocodile (C. porosus), and the Orinoco crocodile (C. intermedius)—have been known to attack and eat humans. Presumably, the great carnivorous dinosaurs—such as Allosaurus and

Tyrannosaurus—devoured even larger prey. These predators were almost certainly capable of killing the largest of their herbivorous contemporaries.

Locomotion

Walking and Crawling

In the typical reptilian posture, limbs project nearly perpendicular from the body and bend downward toward the ground at the elbows and knees. This limb posture produces a sprawled gait that some biologists label as inefficient and awkward. Its continued persistence in thousands of amphibians and reptiles shows its effectiveness and high efficiency for lifestyles designed for energy conservation. At rest the reptilian trunk and tail lie on the substrate; during walking and running the body is held only slightly above the substrate and bends from side to side to increase the length of each step from each sprawled limb. A few terrestrial reptile groups exhibit an evolutionary shift in limb posture from the horizontal to the vertical. This same shift produces the erect posture seen today in birds and mammals. This vertical posture was typical of late dinosaurs, and presumably, like those of birds and mammals, the dinosaur joints had locking mechanisms to reduce the muscle energy required to hold the body erect when standing still.

Spotted racerunner (Cnemidophorus sacki).

The only living reptiles that use a vertical limb posture in walking are the crocodiles. The "high walk" of these animals employs the quadrupedal limb-movement sequence with only a slight lateral undulation. Some young crocodiles use a galloping gait, much like that of a bounding rabbit, for high-speed escape; the body flexes up and down rather than from side to side for this type of locomotion. Crocodiles also use a belly-slide gait. This is also an escape behaviour, even though the body rests on the substrate; lateral undulations of the body and tail and the quadrupedal limb sequence propel the crocodile down the river bank into the water.

A snake moves by pushing backward against rocks, sticks, or any relatively fixed point—such as a lump of earth or a small depression in uneven ground—with the rear (ventrolateral) surface of the body's curves. Each joint of the body passes through the same curves, pressing against the same object and thrusting forward. Heavy-bodied snakes such as pythons and certain rattlesnakes can move forward without undulation. This rectilinear movement depends on the ability of snakes to stretch or contract their bodies along its longitudinal axis. By raising a part of its belly, stretching that part forward, lowering it to the ground, and repeating the process alternately with other parts of the body, a heavy snake can move forward smoothly in a straight line.

A variety of modern lizards are bipedal when running. The collared lizard (Crotaphytus collaris) of the United States and the frilled lizard (Chlamydosaurus kingii) of Australia are capable of bipedal movement, a phenomenon that was widespread among the dinosaurs. These present-day lizards run on their long hind limbs, with the forward parts of their bodies at an angle of about 60° off the horizontal.

Presumably, bipedalism among the dinosaurs began, as it did among modern lizards, as a means of obtaining a faster running speed. Because the centre of gravity is in front of the hips, modern bipedal lizards must move forward continuously in order to maintain a semierect posture; they can stand still in that position only for very short periods.

The sprawled limb posture of bipedal lizards causes each limb to swing outward as it is brought forward and to push the body sideways and forward when each leg thrusts backward against the ground. Bipedal dinosaurs eliminated this side-to-side motion by shifting to a vertical hind-limb posture. This posture supports the body in an upright position, and the limbs swing directly forward and backward. So successful was this mode of locomotion that dinosaurs utilizing it dominated terrestrial life for millions of years.

Clinging and Climbing

Arboreal animals possess groups of anatomical features that help them cling to branches and other substrates. The most common clinging structures in vertebrates are claws; they seem to be the only arboreal adaptations of some lizards, such as the common iguana (I. iguana). Similar structures appear in many geckos (family Gekkonidae), in the anoles (Anolis; family Iguanidae), and in some skinks (family Scincidae).

Other adaptations for climbing include footpads. Pads on the feet consist of wide plates or scales under the fingers and toes. The outer layer of each scale is composed of numerous microscopic hooks formed by the free, bent tips of cells. These minute hooks can catch in the slightest irregularities of a surface, and they enable geckos to run up apparently smooth walls and even upside down on plaster ceilings. Because the hook-like cells are bent downward and to the rear, a gecko must curl its pads upward to disengage them. Thus, when walking or running up a tree or wall, a gecko must curl and uncurl its pad surface with every step.

The giant Solomon Islands skink (Corucia), true chameleons (Chamaeleonidae), arboreal vipers, boas, and pythons use prehensile tails—that is, tails that are capable of supporting most of the weight of the animal or are used habitually for grasping—for clinging to their aerial supports. Still, true chameleons rely mainly on the tonglike arrangement of the digits in their feet. The toes of each foot are united into two opposed bundles—three on the inside and two on the outside of the forefoot, and two on the inside and three on the outside of the hind foot.

Slender vine snakes of several genera of the family Colubridae are capable of extending half the body length in a horizontal plane without support; they do so habitually in bridging the gap between tree branches. Most snakes can reach across an open space. However, all except the vine snakes can extend only a short distance, and that portion of the body invariably sags like a cable. In contrast, the vine snakes bridge an open space like an I-beam. This ability is based partly on their reduced body weight and partly on their deepened and strengthened vertebrae.

Swimming

In water, of course, limb movements—whether bipedal or quadrupedal—that work well in terrestrial environments are not very effective. Aquatic reptiles, with some exceptions, use the same means of propulsion as do fish—that is, lateral undulations of the rear half of the body and tail. Crocodiles and aquatic lizards, such as some monitors (family Varanidae) and the marine iguana (Amblyrhynchus cristatus), undulate their bodies and tails from side to side while holding the limbs against the body. The ancient mesosaurs (order Mesosauria) and ichthyosaurs (order Ichthyosauria) used the same method. The marine ichthyosaurs, which were the reptilian counterpart of the porpoises (family Phocoenidae) in class Mammalia, may have used their flippers as rudders.

The limbless snakes are good swimmers and make lateral undulations similar to those of eels. This fishlike swimming mode requires a flexible body and, usually, a tail of moderate length. Sea snakes have laterally flattened tails that increase their locomotor power. Turtles propel themselves by using their feet as paddles as a part of their quadrupedal limb-movement sequence. Freshwater turtles have webbed feet, whereas the forelimbs of marine turtles are essentially flattened flippers that are moved in a sweeping figure-eight pattern through the water.

Flying

Several groups of reptiles have experimented with flight. One group within the Archosauria (the "ruling reptiles" that include dinosaurs and crocodiles) became highly successful at this means of locomotion and evolved into birds.

Another group of archosaurs, the Pterosauria, developed wings that were supported along the front margin by the arm and an extremely elongated finger. The pterosaur wing was made of skin; since it lacked both internal supports and feathers, it probably lacked the flexibility or durability of a bird wing. Pterosaurs seem to have emphasized soaring and gliding during flight, but they also engaged in flapping flight. It is possible that pterosaurs had clumsy takeoffs like those of the modern albatross (Diomedea).

Since most pterosaur remains have been found in marine deposits, it is assumed that many of the species lived along ocean shores, probably roosting on cliffs from which takeoff would have been easier.

Among modern lizards, flying lizards (Draco) are expert gliders. The “wing” of these small reptiles is made up of skin supported by five or six elongated ribs between the forelimbs and hind limbs. At rest the ribs and wings are folded against the sides of the body. During flight the wings form broad semicircles on each side between the limbs. These lizards, which live in the forested country of Southeast Asia and the East Indies, are gliders; in spite of their label as flying lizards, they are not flyers. To glide, the lizard launches itself from a tree into the air toward another tree, turning upward sharply to slow its body just before lighting on the new perch. Since the limbs are not modified, this lizard walks and runs like any other arboreal lizard.

Form and Function

External Covering

The external covering of reptiles is dry and composed of scales made of keratin. It bears few glands or none at all and differs in this respect from the skin of amphibians and mammals. The epidermis has cycles of growth, and in the outer layer the cells die and harden into a tough and horny surface. Bony plates (osteoderms) develop in the dermis, the layer just below the epidermis, of some reptiles and are best seen on the backs of crocodiles.

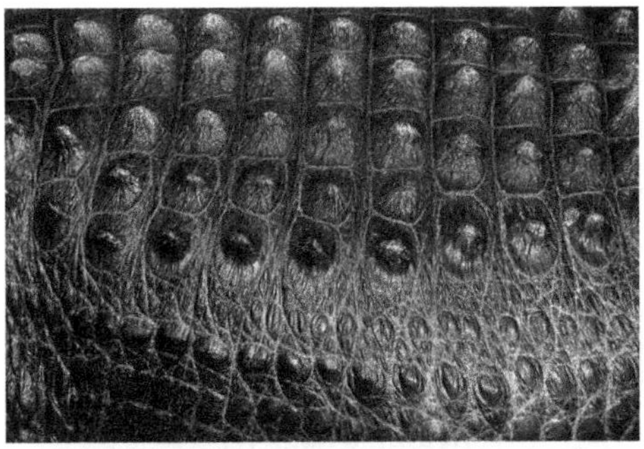

Alligator skin close-up of alligator skin.

Internal Features

Skeletal System

The skeletons of reptiles fit the general pattern of vertebrates. They have a bony skull, a long vertebral column that encloses the spinal nerve cord, ribs that form a protective bony basket around the viscera, and a framework of limbs.

Each group of reptiles developed its own particular variations on this major pattern in accord with the general adaptive trends of the group. Snakes, for example, have lost the limb bones, although a few retain vestiges of the hind limbs. The limbs of several types of marine reptiles have been modified into fins or flippers. In other types, such as the extinct marine-dwelling ichthyosaurs and plesiosaurs, the bones of the limbs, which no longer needed to support the weight of the body against the pull of gravity, became much shorter. At the same time, the bones of other reptiles that composed the digits multiplied in number, forming a long flipper.

Groups of reptiles whose modes of life came to depend heavily on passive defense also developed specializations of the skeleton. The bony and horny shells of turtles and the rows of bony plates on the backs of crocodiles and the Ankylosaurus (a genus of dinosaurs that lived between the Early Jurassic Period and the end of the Cretaceous) are cases in point.

Skull and Dentition

The skulls of the several subclasses and orders vary in the ways mentioned below. In addition to differences in openings on the side of the skull and in general shape and size, the most significant variations in reptilian skulls are those affecting movements within the skull.

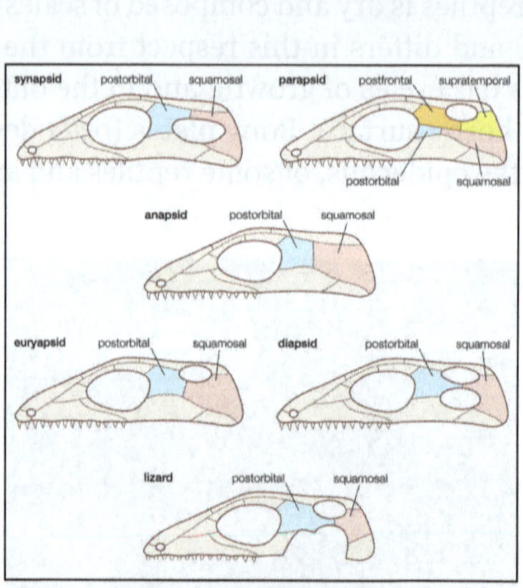

Reptilian skull types.

As a group, reptilian skulls differ from those of early amphibians. Reptiles lack an otic notch (an indentation at the rear of the skull) and several small bones at the rear of the skull roof. The skulls of modern reptiles are also sharply set off from those of mammals in many ways, but the clearest differences occur in the lower jaw and adjacent regions. Reptiles have a number of bones in the lower jaw, only one of which, the dentary, bears teeth. Behind the dentary a small bone, the articular, forms a joint with the quadrate bone near the rear of the skull. In contrast, the lower jaw of a mammal is made up of a single bone, the dentary; the articular and quadrate have become part of the chain of little bones in the middle ear. An almost complete transition between these two very different arrangements is known from fossils of early synapsids (order Therapsida).

The dentition of most reptiles shows little specialization in a given row of teeth. A dentition that divides groups of teeth into distinctive bladelike incisors, tusklike canines, and flat-crowned molars occurs in mammals but does not occur in reptiles. Instead, the entire tooth row is usually made up of long conical teeth. Venomous snakes have one or several hollow or grooved fangs, but they have the same shape as most snake teeth. The principal differences between species lie in the number, length, and position of the teeth. Crocodiles among the living forms and dinosaurs among the extinct forms have but a single upper and a single lower tooth row. Snakes and many extinct reptilian groups have teeth on the palatal bones (vomer, palatine, pterygoid) and on the bones of the upper jaw (premaxilla, maxilla). However, only one row of teeth is present on the lower jaw.

Lizards have conical or bladelike bicuspid or tricuspid teeth. Some species have conical teeth at the front of the jaws and cuspid teeth toward the rear, but the latter are not comparable to the molars of mammals in either form or function. (They are neither flat-crowned nor used to grind food.) Turtles, except for the earliest extinct species, lack teeth. Instead, they have upper and lower horny plates that serve to bite off chunks of food.

The teeth of reptiles are also less specialized in function than are mammalian teeth. The larger carnivorous reptiles are equipped only to tear off or bite off large pieces of their prey and swallow them without chewing. Insectivorous lizards, which constitute the majority of all lizards, usually crack the exoskeleton of their insect prey, and then they swallow the prey without grinding it up. Snakes simply swallow their prey whole without any mechanical reduction, although the puncture wounds permit digestive enzymes to enter the prey to aid digestion.

Many reptiles developed joints (in addition to the hinge for the lower jaw) within the skull that permit the slight movement of one part relative to others. The capacity for such movement within the skull, called kinesis, enables an animal to increase the gape of the mouth and thus is an adaptation for swallowing large objects. Apparently some of the large carnivorous theropod dinosaurs (such as Allosaurus) had a joint between the frontal and parietal bones in the roof of the skull. All reptiles of the super order Lepidosauria (lizards, snakes, and tuatara) have kinetic skulls, but they differ from the dinosaurs in that the joint on the floor of the skull occurs at the juncture of basisphenoid and pterygoid bones in lepidosaurians.

The skulls of the lepidosaurians became increasingly kinetic as new groups evolved. The Sphenodontia (which include living tuatara) and their antecedents, the Rhynchocephalia, had only the basisphenoidal-pterygoidal joint. The lizards lost the lower temporal bar, thereby freeing the quadrate bone and allowing greater movement to the lower jaw, which is hinged to the quadrate. Finally, in the snakes, this trend culminates in the most kinetic skull among the vertebrates. The skulls of snakes possess the ancestral basisphenoidal-pterygoidal joint, a highly mobile quadrate (which gives even greater mobility to the lower jaw), and upper jaws capable of rotating on their longitudinal axes and moving both forward and backward. Many snake species also have a hinge on the roof of the skull between the nasal and frontal bones that allows the snout to be raised slightly. In short, the only part of a snake's skull incapable of movement is the braincase.

Nervous System

As in all vertebrates, the nervous system of reptiles consists of a brain, a spinal nerve cord, nerves running from the brain or spinal cord, and sense organs. When compared with mammals, reptiles have proportionately smaller brains. The most important difference between the brains of these two vertebrate groups lies in the size of the cerebral hemispheres, the principal associative centres of the brain. These hemispheres make up the bulk of the brain in mammals and, when viewed from above, almost hide the rest of the brain. In reptiles the relative and absolute size of the cerebral hemispheres is much smaller. The brain of snakes and alligators forms less than 1/1,500 of total body weight, whereas in mammals such as squirrels and cats the brain accounts for about 1/100 of body weight.

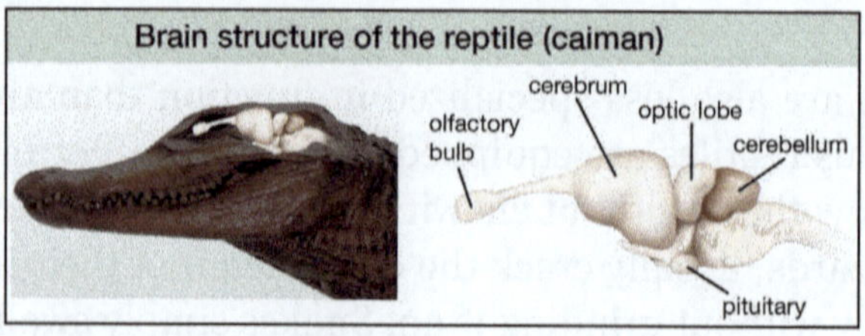

In the brain of reptiles such as the caiman (related to the alligator and crocodile), the olfactory bulb (regulating the sense of smell) remains a large and important structure.

Circulatory System

Modern reptiles do not have the capacity for the rapid sustained activity found in birds and mammals. It is generally accepted that this lower capacity is related to differences in the circulatory and respiratory systems. Before the origin of lungs, the vertebrate circulatory system had a single circuit: in the fishes, blood flows from heart to gills to body and back to the heart. The heart consists of four chambers arranged in a linear sequence.

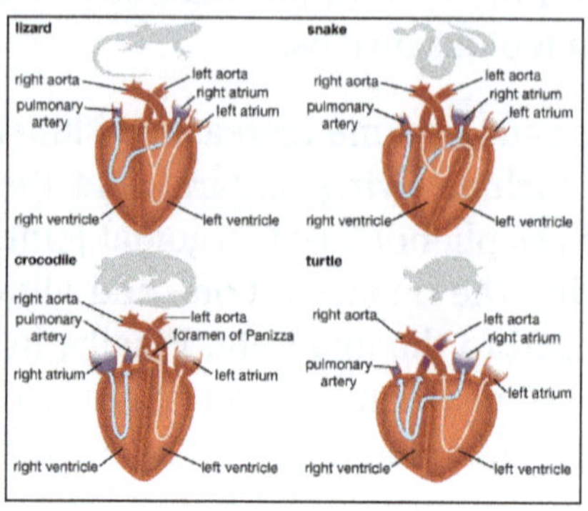

Types of reptilian hearts: lizard, snake, crocodile, and turtle.

With the evolution of lungs in early tetrapods, a new and apparently more efficient circulatory system evolved. Two chambers of the heart, the atrium and ventricle, became increasingly important, and the beginnings of double circulation appeared. An early stage in this evolution can be seen in amphibians today, where pulmonary artery from the heart the goes directly to the lungs, whereas the systemic arteries carry blood to the general body. In amphibians the blood is aerated in the lungs and carried back to the atrium of the heart. From the left side of the atrium, which is at least partially divided, the aerated blood is pumped into the ventricle to mix with nonaerated blood; nonaerated blood from the body is returned to the heart via the right half of the atrium. Then the cycle begins again. One aspect of the amphibian system is that the blood leaving the heart for the body is only partially aerated; part of it is made up of deoxygenated blood returned from the body.

All groups of modern reptiles have a completely divided atrium; it is safe to assume, therefore, that this was true of most, if not all, extinct reptiles. In the four major living groups of reptiles, the ventricle is at least partially divided. When the two atria of a lizard's heart contract, the two streams of blood (aerated blood from the lungs in the left atrium and nonaerated blood from the body in the right atrium) flow into the left chamber of the ventricle. As pressure builds up in that chamber, the nonaerated blood is forced through the gap in the partition into the right chamber of the ventricle. Then, when the ventricle contracts, nonaerated blood is pumped into the pulmonary artery and thence to the lungs, whereas aerated blood is pumped into the systemic arteries (the aortas) and so to the body.

In snakes all three arterial trunks come out of the chamber of the ventricle that receives the nonaerated blood of the right atrium. During ventricular contraction a muscular ridge forms a partition that guides the nonaerated blood into the pulmonary artery, while the aerated blood received by the other chamber of the ventricle is forced through the opening in the ventricular septum and out through the aortas.

In crocodiles the ventricular septum is complete, but the two aortas come out of different ventricular chambers. A semilunar valve at the entrance to the left aorta prevents nonaerated blood in the right ventricle from flowing into the aorta. Instead, part of the aerated blood from the left ventricular chamber pumped into the right aorta flows into the left by way of an opening.

The ventricle of the turtle is not perfectly divided, and some slight mixing of aerated and nonaerated blood can occur.

Despite the peculiar and complex circulation, lizards, snakes, and crocodilians have achieved a double system. Tests of the blood in the various chambers and arteries have shown that the oxygen content in both systemic aortas is as high as that of the blood just received by the left atrium from the lungs and is much higher than that of the blood in the pulmonary artery. Except for the turtles, limitation of activity in reptiles cannot be explained on the basis of heart circulation. An explanation may lie in the chemistry of the blood. The blood of reptiles has less hemoglobin and thus carries less oxygen than that of mammals.

Respiratory System

The form of the lungs and the methods of irrigating them may also influence activity by affecting the efficiency of gas exchange. In snakes the lungs are simple saclike structures having small pockets, or alveoli, in the walls. In the lungs of all crocodiles and many lizards and turtles, the surface area is increased by the development of partitions that, in turn, have alveoli. Because exchange of respiratory gases takes place across surfaces, an increase of the ratio of surface area to volume leads to an increase in respiratory efficiency. In this regard the lungs of snakes are not as effective as the lungs of crocodiles. The elaboration of the internal surface of lungs in reptiles is simple, however, compared with that reached by mammalian lungs, with their enormous number of very fine alveoli.

Most reptiles breathe by changing the volume of the body cavity. By contractions of the muscles moving the ribs, the volume of the body cavity is increased, creating a negative pressure, which is restored to atmospheric level by air rushing into the lungs. By contraction of body muscles, the volume of the body cavity is reduced, forcing air out of the lungs.

This system applies to all modern reptiles except turtles, which, because of the fusion of the ribs with a rigid shell, are unable to breathe by this means; they do use the same mechanical principle of changing pressure in the body cavity, however. Contraction of two flank muscles enlarges the body cavity, causing inspiration. Contraction of two other muscles, coincident with relaxation of the first two, forces the viscera upward against the lungs, causing exhalation.

The rate of respiration, like so many physiological activities of reptiles, is highly variable, depending in part upon the temperature of the environment and in part upon the emotional state of the animal.

Digestive and Urogenital Systems

The digestive system of modern reptiles is similar in general plan to that of all higher vertebrates. It includes the mouth and its salivary glands, the esophagus, the stomach, and the intestine and ends in a cloaca. Of the few specializations of the reptilian digestive system, the evolution of one pair of salivary glands into poison glands in the venomous snakes is the most remarkable.

During development the embryos of higher vertebrates (reptiles, birds, and mammals) consecutively develop three separate sets of kidneys; these are arranged in longitudinal sequence in the body cavity. The first set, the pronephroi, are vestigial organs left over from the evolutionary past that soon degenerate and disappear without having had any function. The second set, the mesonephroi, are the functional kidneys of adult amphibians, but their only contribution to the lives of reptiles is in providing the duct (the Wolffian duct) that forms a connection between the testes and the cloaca. The operational kidneys of reptiles, birds, and mammals are the last set, the metanephroi, which have separate ducts to the cloaca. The principal functions of the kidney are the removal of nitrogenous wastes resulting from the oxidation of proteins and the regulation of water loss. Vertebrates eliminate three kinds of nitrogenous wastes: ammonia, urea, and uric acid. Ammonia and urea are highly soluble in water, but uric acid is not. Ammonia is highly poisonous, urea is slightly poisonous, and uric acid is not poisonous at all.

Among reptiles the form taken by the nitrogenous wastes is closely related to the habits and habitat of the animal. Aquatic reptiles tend to excrete a large proportion of these wastes as ammonia in aqueous solution. This method uses large amounts of water and is no problem for a freshwater resident, such as an alligator, which eliminates between 40 and 75 percent of its nitrogenous wastes as ammonia. Terrestrial reptiles, such as most snakes and lizards, must conserve body water, and they convert their nitrogenous wastes to insoluble, harmless uric acid, which forms a more or less solid mass in the cloaca. In snakes and lizards, these wastes are eliminated from the cloaca together with wastes from the digestive system.

Prior to the evolution of the metanephric kidney, the products of the male gonad, the testis, traveled through the same duct with the nitrogenous wastes from the kidney. But with the appearance of the metanephros, the two systems became separated. The female reproductive system never shared a common tube with the kidney. Oviducts in all female vertebrates arise as separate tubes with openings usually near, but not connected to, the ovaries. The oviducts, like the Wolffian ducts of the testes, open to the cloaca. Both ovaries and testes lie in the body cavity near the kidneys.

With the evolution of the reptilian egg, internal fertilization became necessary. The males of all modern reptiles, with the exception of tuatara, have functional copulatory organs. The structures vary from group to group, but all include erectile tissue as an important element of the operating mechanism, and all are protruded through the male's cloaca into that of the female during copulation. Unlike the penis of turtles and crocodiles, the copulatory organ of lizards and snakes is paired, with each unit being called a hemipenis. The hemipenes of lizards and snakes are elongated tubular structures lying in the tail. The penis of a crocodile or turtle is protruded through the cloacal opening wholly by means of a filling of blood space (sinuses) in the penis; protrusion of a lizard's or snake's hemipenis, however, is begun by a pair of propulsor muscles. Completion of the erection is brought about by blood filling the sinuses in the erectile tissue. Only one hemipenis is inserted into a female, but which one is a matter of chance. Unlike the penis of mammals, the copulatory organs of reptiles do not transport sperm through a tube. The ducts from the testes, as already mentioned, empty into the cloaca, and the sperm flow along a groove on the surface of the penis or hemipenis.

Sense organs

Sight

In general construction the eyes of reptiles are like those of other vertebrates. Accommodation for near vision in all living reptiles except snakes is accomplished by pressure being exerted on the lens by the surrounding muscular ring (ciliary body), which thus makes the lens more spherical. In snakes the same end is achieved by the lens being brought forward. The lens moves as a result of the pressure built up on the vitreous humour by contractions of muscles located at the base of the iris. The pupil shape varies remarkably among living reptiles, from the round opening characteristic of all turtles and many diurnal lizards and snakes to the vertical slit of crocodiles and nocturnal snakes and the horizontal slits of a few tree snakes. Undoubtedly the most bizarre pupil shape is that of some geckos, in which the pupil contracts to form a series of pinholes, one above the other. The lower eyelid has the greater range of movement in most reptiles. In crocodiles the upper lid is more mobile. Snakes have no movable eyelids, their eyes being covered by a fixed transparent scale. tuatara and all crocodiles have a third eyelid, the nictitating membrane, a transparent sheet that moves sideways across the eye from the inner corner, cleansing and moistening the cornea without shutting out the light.

Specialized eyes of the chameleon (Chamaeleo) and the gecko (Gekko).

Visual acuity varies greatly among living reptiles, being poorest in the burrowing lizards and snakes (which often have very small eyes) and greatest in active diurnal species (which usually have large eyes). Judging by the size of the skull opening in which the eye is situated, similar variation existed among the extinct reptiles. Extinct forms, such as the ichthyosaurs, that hunted active prey had large eyes and presumably excellent vision; many herbivorous types, such as the horned dinosaur Triceratops, had relatively small eyes and weak vision. Colour vision has been demonstrated in few living reptiles.

Hearing

The power of hearing is variously developed among living reptiles. Crocodiles and most lizards hear reasonably well. Snakes and turtles are sensitive to low-frequency vibrations, thus they "hear" mostly earth-borne, rather than aerial, sound waves. The reptilian auditory apparatus is typically made up of a tympanum, a thin membrane located at the rear of the head; the stapes, a small bone running between the tympanum and the skull in the tympanic cavity (the middle ear); the inner ear; and a eustachian tube connecting the middle ear with the mouth cavity. In reptiles that can hear, the tympanum vibrates in response to sound waves and transmits the vibrations to the stapes. The inner end of the stapes abuts against a small opening (the foramen ovale) to the cavity in the skull containing the inner ear. The inner ear consists of a series of hollow interconnected parts: the semicircular canals; the ovoidal or spheroidal chambers called the utriculus and sacculus; and the lagena, a small outgrowth of the sacculus. The tubes of the inner ear, suspended in a fluid called perilymph, contain another fluid, the endolymph. When the stapes is set in motion by the tympanum, it develops vibrations in the fluid of the inner ear; these vibrations activate cells in the lagena, the seat of the sense of hearing. The semicircular canals are concerned with equilibrium.

Most lizards can hear. The majority have their best hearing in the range of 400 to 1,500 hertz and possess a tympanum, a tympanic cavity, and a eustachian tube. The tympanum, usually exposed at the surface of the head or at the end of a short open tube, may be covered by scales or may be absent. In general the last two conditions are characteristic of lizards that lead a more or less completely subterranean life. For subterranean lizards airborne sounds are less important than the low-frequency sounds passing through the ground. The middle ear of these burrowers is usually degenerate as well, often lacking the tympanic cavity and eustachian tube.

Snakes have neither tympanum nor eustachian tube, and the stapes is attached to the quadrate bone on which the lower jaw swings. Snakes are obviously more sensitive to vibrations in the ground than to airborne sounds. A loud sound above a snake does not elicit any response, provided that the object making the sound does not move or, if it does, the movements are not seen by the snake. On the other hand, the same snake will raise its head slightly and flick its tongue in and out rapidly if the ground behind it is tapped or scratched. Snakes undoubtedly "hear" these vibrations by means of bone conduction. Sound waves travel more rapidly and strongly in solids than in the air and are probably transmitted first to the inner ear of snakes through the lower jaw, which is normally touching the ground, thence to the quadrate bone, and finally to the stapes. Burrowing lizards presumably hear ground vibrations in the same way.

All crocodiles have rather keen hearing and have an external ear made up of a short tube closed by a strong valvular flap that ends at the tympanum. The American alligator (Alligator mississippiensis) can hear sounds within a range of 50 to 4,000 hertz. The hearing of crocodiles is involved not only in the detection of prey and enemies but also in their social behaviour; males roar or bellow to either threaten other males or to attract females.

have well-developed middle ears and usually large tympana. Measurements of the impulses of the auditory nerve between the inner ear and the auditory centre of the brain show that the inner ear in several species of turtles is sensitive to airborne sounds in the range of 50 to 2,000 hertz.

Chemoreception

Chemically sensitive organs, used by many reptiles to find their prey, are located in the nose and in the roof of the mouth. Part of the lining of the nose is made up of cells subserving the function of smell and corresponding to similar cells in other vertebrates. The second chemoreceptor is the Jacobson's organ, which originated as an outpocketing of the nasal sac in amphibians; it remained as such in tuatara and crocodiles. The Jacobson's organ is most developed in lizards and snakes, in which its connection with the nasal cavity has been closed and is replaced by an opening into the mouth. The nerve connecting Jacobson's organ to the brain is a branch of the olfactory nerve. In turtles the Jacobson's organ has been lost.

A black-and-yellow mangrove snake (Boiga dendrophila) sticking out its forked tongue. A snake uses its tongue to deliver heavy airborne odour particles to its Jacobson organ.

The use of the Jacobson's organ is most obvious in snakes. If a strong odour or vibration stimulates a snake, its tongue is flicked in and out rapidly. With each retraction, the forked tip touches the roof of the mouth near the opening of the Jacobson's organ, transferring any odour particles adhering to the tongue. In effect, the Jacobson's organ is a short-range chemoreceptor of nonairborne odours, as contrasted to the detection of airborne odours, smelling in the usual sense, by olfactory sensory patches in the nasal tube.

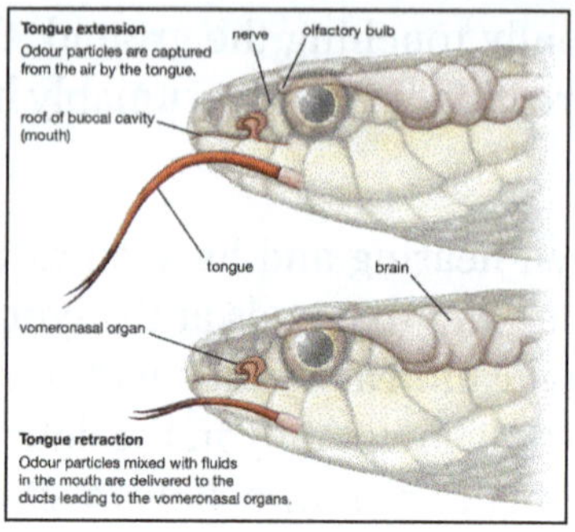

The process of chemoreception using the Jacobson's, or vomeronasal, organ.

Some snakes (notably the large vipers) and scleroglossan lizards (such as skinks, monitors, and burrowing species of other families) rely upon the olfactory tissue and the Jacobson's organ to locate food, almost to the exclusion of other senses. Other reptiles, such as certain diurnal lizards and crocodiles, appear not to use scent in searching for prey, though they may use their sense of smell for locating a mate.

The pit vipers (family Viperidae), boas and pythons (family Boidae), and a few other snakes have special heat-sensitive organs (infrared receptors) on their heads as part of their food-detecting apparatus. Just below and behind the nostril of a pit viper is the pit that gives the group its common name. The lip scales of many pythons and boas have depressions (labial pits) that are analogous to the viper's pit. The labial pits of pythons and boas are lined with skin thinner than that covering the rest of the head and are supplied with dense networks of blood capillaries and nerve fibres. The facial pit of the viper is relatively deeper than the boa's labial pits and consists of two chambers separated by a thin membrane bearing a rich supply of fine blood vessels and nerves. In experiments using warm and cold covered electric light bulbs, pit vipers and pitted boas have been shown to detect temperature differences of less than 0.6 °C (1.1 °F).

Many pit vipers, pythons, and boas are nocturnal and feed largely on mammals and birds. Infrared receptors, located on the face, enable these reptiles to direct their strikes accurately in the dark, once their warm-blooded prey arrives within range. The approach of prey is likely identified by the vibrations they make on the ground; however, the sense of sight and perhaps even the sense of smell are also used. The pit organs simply confirm the identity of the prey and aim the strike.

Thermal Relationships

Reptiles are often described as being cold-blooded animals. However this is not always true. They have no internal mechanism for the production of heat and maintenance of an elevated body temperature; they are dependent upon heat from their surroundings; that is, they are ectothermic. As ectotherms, many reptiles have body temperatures which fluctuate with that of the environment. This condition is called poikilothermy. Mammals and birds, often described as warm-blooded animals, produce heat by a cellular process and maintain relatively high body temperatures independent of the environment. In mammals, body temperature is kept relatively constant, and this condition is termed homoiothermy. For example, when the body temperature of a dog or a human being falls below the normal range, shivering begins, and blood vessels in the skin contract. Subsequent muscular activity generates heat, and the contraction of the superficial blood vessels, by reducing the volume of blood flow at the surface, reduces heat loss by radiation. By contrast, when the body temperature of a reptile falls below the optimum, it must move to a part of the environment with a higher ambient temperature. When environmental temperatures fall below a critical minimum, a reptile's metabolic activity decreases; its movements become sluggish, its heartbeat slows, and its rate of breathing drops. In short, it becomes incapable of the normal activities required for growth, reproduction, and survival.

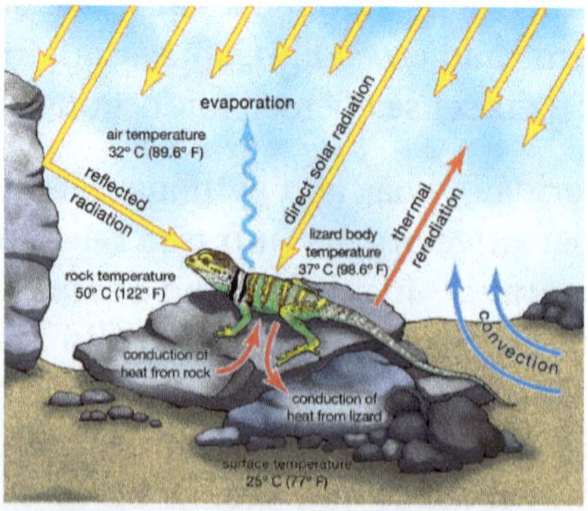

Reptiles are common faunal inhabitants of scrubland ecosystems. Because they are ectotherms, that is, organisms that depend on external sources of heat to regulate body temperature, they use the complex temperature profile of the scrubland environment to derive warmth.

In higher-temperature environments mammals and birds have some physiological means of cooling their bodies. They can pant or sweat, and superficial blood vessels may expand. However, a reptile must ordinarily move away from a spot in which the temperature is too high, or it will perish very quickly. Some reptiles also pant, but their temperature accommodations are largely behavioral; they might change their orientation with respect to the sun or wind or raise their body from the ground.

Each group of reptiles has its own characteristic thermal range. One genus of lizards, for example, may require temperatures of 29–32 °C (84–90 °F) for maximum efficiency, whereas another may require temperatures of 24–27 °C (75–81 °F). As a result of such physiological differences, lizards of the two groups will be active at different times of the day or occupy slightly different habitats.

In reptiles the body temperatures at which normal activities occur are generally lower than those of most mammals. However, a few sun-loving (heliothermic) lizards, such as the greater earless lizard (Holbrookia texana) of the southwestern United States, have average activity temperatures above 38 °C (100 °F). This temperature is slightly higher than the average human body temperature. Such high temperatures are exceptional, and the majority of lizards have normal activity temperatures in the 27–35 °C (81–95 °F) range.

Snake

Snake is any of the numerous elongate, limbless, scaled, carnivorous reptiles comprising the suborder Serpentes (or Ophidia) of the order Squamata. Snakes generally are distinguished from lizards—which also belong to order Squamata, but are placed in suborder Sauria (or Lacertilia)— by the absence of legs, although some lizards are legless. Snakes are also differentiated from lizards by have more flexible jaws and by lacking external ear openings and movable eyelids whereas most lizards have movable eyelids and external ear openings.

The more than 2,900 species of snakes are spread across every continent except Antarctica. They range in size from the tiny thread snake to pythons and anacondas over 10 meters long.

External Features

1. The body of snakes is extremely elongated and cylindrical.
2. The limbs are absent, except in certain genera where traces of posterior limbs are present at the base of the tail and bear conical claws.

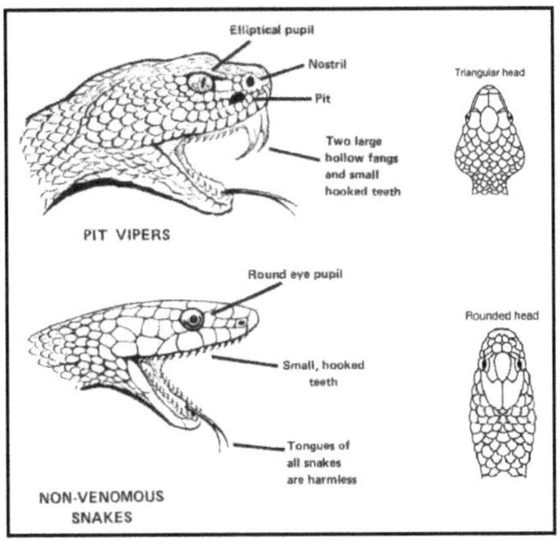

3. Distinct neck is absent.
4. The body is covered with scales and shields.
5. The osteoderms are absent.
6. The scales are usually small and overlap each other, but the shields are usually large and join each other by their margins.
7. The head is covered mostly by the head shields.
8. The head shields are named according to their placement.
9. The shields situated in the mental groove on the ventral side between the rami of the lower jaws are called mental shields. The mental shields are considered to be peculiar features.
10. The other shields are frontal shield, super-ciliary shields, rostral shield, upper labial shields, nasal shield, pre-orbital shields, loreal shield, median labial shield, jugular shields, ventral shields, etc.
11. The scales on the ventral side of the body are broad and cover the body like transverse bands. These ventral scales are called gastrosteges.

12. The whole of the scaly epidermis is cast off several times in a year. This phenomenon is called the ecdysis or moulting.
13. The head is more or less conical.
14. The nostrils are fused over the eyes.
15. The tympanic membrane, tympanic cavity and Eustachian tube are absent.
16. The cloacal aperture is a transverse slit.
17. Males are provided with paired copulatory organs.

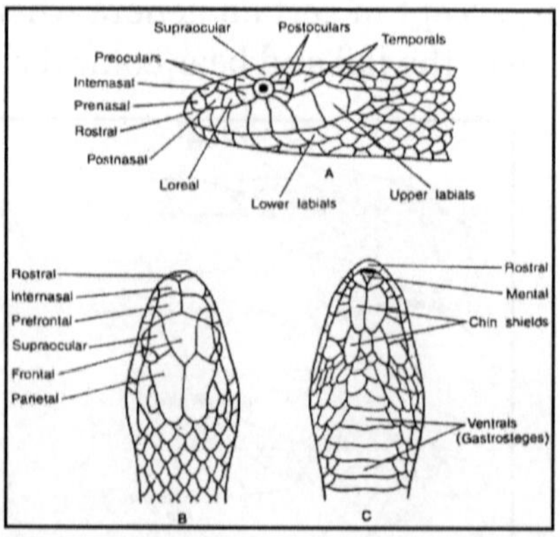

Figure: Different scales showing on the head of the snakes. A Lateral view, B. Doesal View, C. Ventral View

Internal Structures

Soft Parts

The internal peculiar structures of the snakes are seen according to their peculiar mode of living in different habitats.

Digestive System

1. The mouth cavity is bounded by the upper and lower jaws.
2. The tongue is provided with a muscular sheath into which the tongue may be retracted and this part is highly sensitive, and used as a tactile organ.
3. The teeth are conical, backwardly curved and are situated in the maxillae, palatines and pterygoid bones in the jaw. Dentition is pleurodont type.
4. The labial glands are modified into poison glands in case of poisonous snakes and do not help in digestion and the large backwardly directed maxillary teeth become into fangs in venomous snakes.
5. The buccal cavity leads into shorter pharynx.

6. The oesophagus is thin walled and opens into a muscular dilated and elongated part, called stomach.

7. The distal part of the stomach becomes narrower and is marked by the pyloric valve. After the pyloric valve, the intestine begins.

8. The intestine is divided into well- defined small intestine and large intestine that ends in a cloaca.

9. Liver is elongated and is formed by two unequal lobes. The right lobe is longer than the left.

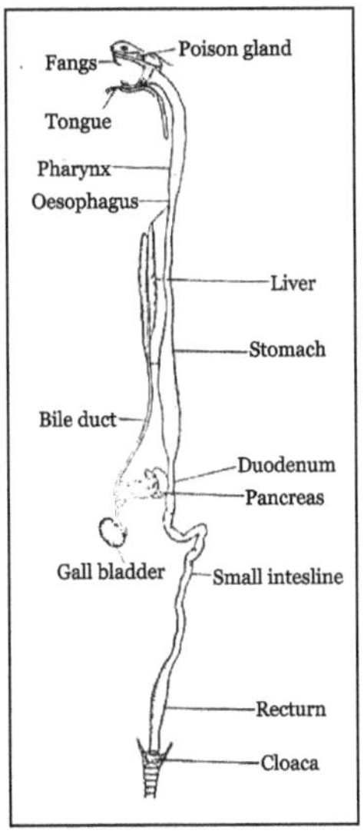

Figure: Digestive system of Saw-Scaled Viper (Echis carinatus).

Respiratory System

1. The respiratory system consists of a long trachea and lungs. The wall of the trachea is supported by numerous cartilaginous rings.

2. The anterior part of the trachea is dilated and form, the larynx which is supported by cricoid and arytenoid cartilages.

3. The lungs in the snakes exhibit an asymmetrical feature and the left lung is often small and sometimes absent.

The right lung is always very long and in marine species, it may extend up to the cloaca and it performs the hydrostatic function. In the more primitive groups both lungs lobes are present (e.g., Python, Boa). The elongated trachea opens at the anterior portion of buccal cavity and be protruded between the two halves of the lower jaw.

This is a modification for respiration during swallowing of prey. In some forms, a part of the trachea is specialized and acts as a tracheal lung. This is due to the incomplete tracheal rings and this part expands as a tracheal membrane. This membrane develops as an alveoli and extends up to the right lung proper.

Circulatory System

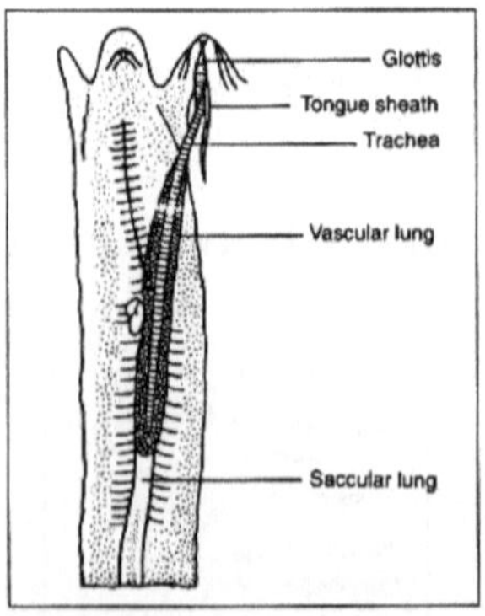

Respiratory system of saw-scaled viper.

1. The heart consists of two auricles and one incompletely divided ventricle. The ventricular septum is well developed which indicates the division of the ventricle.
2. The arterial arches are asymmetrical—mainly the carotid and jugular.
3. Ductus caroticus is absent.

Due to imperfections in the respiratory and circulatory systems, the reptiles, specially the snakes, are compelled to supply enough oxygen to their tissues for food combustion. So their body temperatures can be raised a few degrees above the surrounding temperature.

Nervous System and Organs of Special Sense

1. The brain is narrow, elongated and projected between the eyes and is more highly organised than the amphibians.
2. There are ten pairs of cranial nerves.

Figure: Jacobson's organ is used for taste and smell analysis of the particles that collect from the ground.

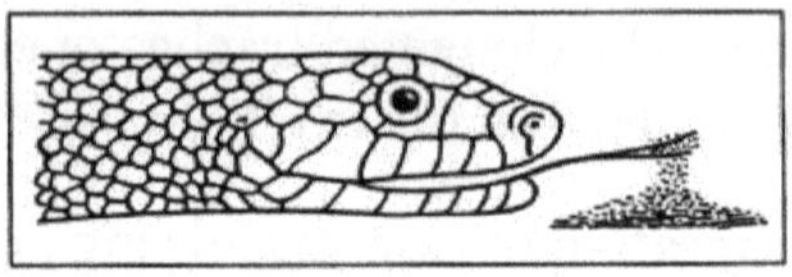

Figure: Bifid tongue is used to pick up particles from the ground.

3. The vomeronasal or Jacobson's organ is well developed and is used for taste and small analysis of the dust particles.
4. The eyes of snakes have many peculiarities (a) There are no scleral ossicles or cartilages and accommodation is brought about by the lens displacement (b) The visual cells include peculiar types of cones, developed from rods. From the above peculiarities of the eyes, Walls concluded that snakes originated as burrowing creatures and later developed as the animals of the above surface of the ground.
5. In the ear portion, the middle ear cavity—the tympanic membranes, tympanic cavity and eustachian tube— is reduced or absent and the columella auris occurs and attached externally to the quadrate. The general belief is that the snakes cannot perceive the air-borne sound.

But according to Marshall and Williams, "the snakes are highly sensitive to earth-borne vibrations due to the attachment of columella to quadrate", and according to Young, "they can hear the air-borne sounds only of low frequency".

Urinogenital System

1. The kidneys are asymmetrical and elongated. In the land snakes the number of glomeruli is highly reduced.
2. The urinary bladder is absent.
3. The gonads are asymmetrical. The males are provided with a pair of eversible copulatory organs, the apex of which bears spiny projections.
4. Anal glands are present at the base of the tail and discharge into cloaca. There is some evidence that the secretion is odorous in nature and attracts the opposite mates during the breeding season.
5. The sexes are distinguished by a number of features:
    - The size of the body—the males are smaller than the females.
    - The sizes of the tail between the male and female sexes. The males have a longer swollen base tail and this swelling is due to the paired hemipenes, the male copulatory organs, at the base of the tail. The tail in the female is generally tapered and ends in a fine point.
6. Snakes are mostly oviparous, some are ovoviviparous or viviparous (laying young ones directly), Uropeltis ocellatus, various vipers (vipera sp.) and sea snakes (except amphibious forms) are ovoviviparous or viviparous in nature.

Oviparous and ovoviviparous can be distinguished in such way as, if laying precedes hatching the animal is called oviparous but the individuals of the same species may hatch eggs in the "oviducts", called ovoviviparous.

Development

1. The eggs are moderately telolecithal and the egg shell is leathery.
2. Segmentation is meroblastic.
3. Some snakes have developed yolk sac or chorioallantoic placentation.
4. The embryos of snakes possess a small "egg tooth" that helps to puncture the shell and shell-membrane.

Hard Part

1. The skull is streptostylic (i.e., the quadrate is highly movable) in nature and occipital condyle is distinctly tripartite.
2. The fossae and inter-parietal foramen are absent.
3. The parietals are fused to form a single piece.
4. The frontals are provided with lateral projections which unite at the floor of the skull.
5. The inter-orbital septum is absent.
6. The palatines are slender and widely separated. They are connected by the pterygoids behind. The transverse or epipterygoid bones are connected with pterygoid and maxillary bones at the rear and front ends.
7. The posterior border of the orbit is formed by the post-frontal.
8. The premaxillae are united into a single piece and are usually toothless. The maxillae, palatines and pterygoids are freely movable.

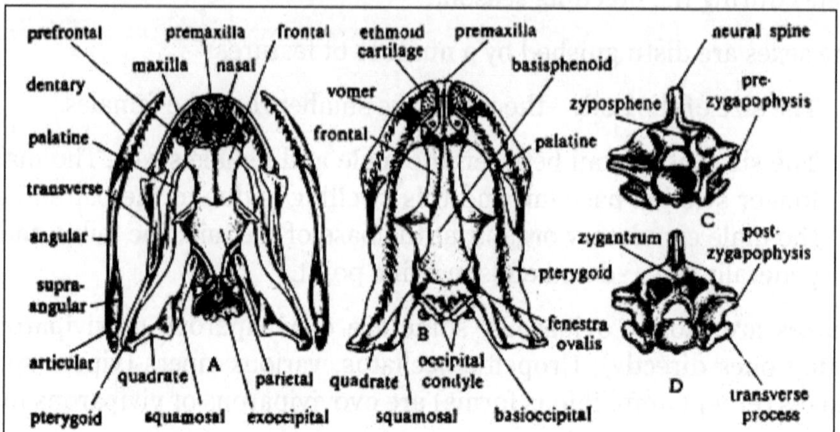

A. Dorsal view of the skull and lower jaw of a non-poisonous snake. B. Ventral view of the skull fo a non-poisonous snake. C. Anterior view and D. Posterior view of a precaudal vertabra of snake.

9. The quadrate and squamosal of the skull are loosely articulated and squamosal is horizontal in position and extends beyond the posterior limit of skull.
10. The rami of the lower jaw are connected by elastic ligamentous tissue so that the gape of the mouth can be increased several times to engulf the large prey during swallowing.

Vertebral Skeleton

1. The vertebrae are strongly procoelous and are divided into precaudal and caudal regions.
2. Besides pre- and post-zygapophyses, extra-articulating surfaces, called zygosphenes, at the anterior face and zygantra at the posterior end are present (Fig. 8.26C, D). The zygosphene is a wedge-like process, situated at the base of the neural arch of the anterior surface.
3. Some of the vertebrae may have a median hypophysis.
4. Transverse processes are rod-shaped and well-developed. These processes at the caudal region are bifid.
5. The ribs are single headed and present in all vertebrae except atlas.
6. Distally the ribs are attached to ventral scales by ligaments to help in locomotion.
7. The girdle and girdle bones are completely absent except Python, only vestige of pelvic girdle is represented.

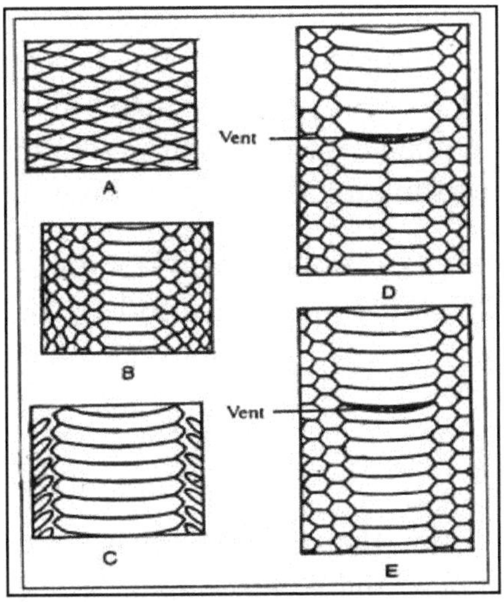

Figure: Scales of different regions of different snakes. A. Body scales (dorsal side) of a primitive snake. B. Scales on the ventral Surface of Phthon. C. Abdominal scales of a cobra. D. Ventral scales around the vent region of a cobra. E. Ventral scales around the vent region of a krait.

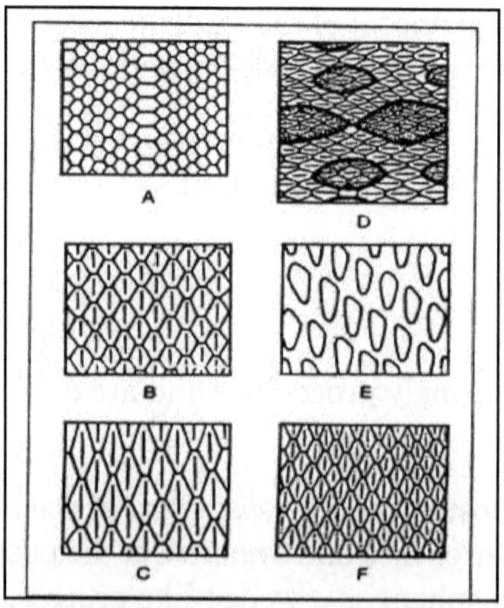

Figure: Scales of the dorsal side of the snakes A. Central row of hexagonal scales (Krait). B. Rat snake (Scales are oblong in shape). C. and D. Russell's viper (Semi-elliptical). E. Rhombold scales in cobra. F. Elliptical Scales in saw scaled viper.

Lizard

Lizards are a widespread group of squamate reptiles, with over 6,000 species, ranging across all continents except Antarctica, as well as most oceanic island chains. The group is paraphyletic as it excludes the snakes and Amphisbaenia; some lizards are more closely related to these two excluded groups than they are to other lizards. Lizards range in size from chameleons and geckos a few centimeters long to the 3 meter long Komodo dragon.

Most lizards are quadrupedal, running with a strong side-to-side motion. Others are legless, and have long snake-like bodies. Some such as the forest-dwelling *Draco* lizards are able to glide. They are often territorial, the males fighting off other males and signalling, often with brightly colours, to attract mates and to intimidate rivals. Lizards are mainly carnivorous, often being sit-and-wait predators; many smaller species eat insects, while the Komodo eats mammals as big as water buffalo.

Lizards make use of a variety of antipredator adaptations, including venom, camouflage, reflex bleeding, and the ability to sacrifice and regrow their tails.

Anatomy

Largest and Smallest

The adult length of species within the suborder ranges from a few centimeters for chameleons such as *Brookesia micra* and geckos such as *Sphaerodactylus ariasae* to nearly 3 m (10 ft) in the case of the largest living varanid lizard, the Komodo dragon. Most lizards are fairly small animals.

Distinguishing Features

Skin of *Lacerta agilis*, showing overlapping scales made of keratin.

Lizards typically have rounded torsos, elevated heads on short necks, four limbs and long tails. Some are legless, including snakes. Lizards and snakes share a movable quadrate bone, distinguishing them from the rhynchocephalians, which have more rigid diapsid skulls . Some lizards such as chameleons have prehensile tails, assisting them in climbing among vegetation.

As in other reptiles, the skin of lizards is covered in overlapping scales made of keratin. This provides protection from the environment and reduces water loss through evaporation. This adaptation enables lizards to thrive in some of the driest deserts on earth. The skin is tough and leathery, and is shed (sloughed) as the animal grows. Unlike snakes which shed the skin in a single piece, lizards slough their skin in several pieces. The scales may be modified into spines for display or protection, and some species have bone osteoderms underneath the scales.

Red tegu (*Tupinambis rufescens*) skull, showing teeth of differing types.

The dentitions of lizards reflect their wide range of diets, including carnivorous, insectivorous, omnivorous, herbivorous, nectivorous, and molluscivorous. Species typically have uniform teeth suited to their diet, but several species have variable teeth, such as cutting teeth in the front of the jaws and crushing teeth in the rear. Most species are pleurodont, though agamids and chameleons are acrodont.

The tongue can be extended outside the mouth, and is often long. In the beaded lizards, whiptails and monitor lizards, the tongue is forked and used mainly or exclusively to sense the environment, continually flicking out to sample the environment, and back to transfer molecules to the vomeronasal organ responsible for chemosensation, analogous to but different from smell or taste. In geckos, the tongue is used to lick the eyes clean: they have no eyelids. Chameleons have very long sticky tongues which can be extended rapidly to catch their insect prey.

Three lineages, the geckos, anoles, and chameleons, have modified the scales under their toes to form adhesive pads, highly prominent in the first two groups. The pads are composed of millions of tiny setae (hair-like structures) which fit closely to the substrate to adhere using van der Waals forces; no liquid adhesive is needed. In addition, the toes of chameleons are divided into two opposed groups on each foot (zygodactyly), enabling them to perch on branches as birds do.

Physiology

Locomotion

Adhesive pads enable geckos to climb vertically.

Aside from legless lizards, most lizards are quadrupedal and move using gaits with alternating movement of the right and left limbs with substantial body bending. This body bending prevents significant respiration during movement, limiting their endurance, in a mechanism called Carrier's constraint. Several species can run bipedally, and a few can prop themselves up on their hindlimbs and tail while stationary. Several small species such as those in the genus *Draco* can glide: some can attain a distance of 60 metres (200 feet), losing 10 metres (33 feet) in height. Some species, like geckos and chameleons, adhere to vertical surfaces including glass and ceilings. Some species, like the common basilisk, can run across water.

Senses

Lizards make use of their senses of sight, touch, olfaction and hearing like other vertebrates. The balance of these varies with the habitat of different species; for instance, skinks that live largely covered by loose soil rely heavily on olfaction and touch, while geckos depend largely on acute vision for their ability to hunt and to evaluate the distance to their prey before striking. Monitor lizards have acute vision, hearing, and olfactory senses. Some lizards make unusual use of their sense organs: chameleons can steer their eyes in different directions, sometimes providing non-overlapping fields of view, such as forwards and backwards at once. Lizards lack external ears, having instead a circular opening in which the tympanic membrane (eardrum) can be seen. Many species rely on hearing for early warning of predators, and flee at the slightest sound.

Nile monitor using its tongue for smell

As in snakes and many mammals, all lizards have a specialised olfactory system, the vomeronasal organ, used to detect pheromones. Monitor lizards transfer scent from the tip of their tongue to the organ; the tongue is used only for this information-gathering purpose, and is not involved in manipulating food.

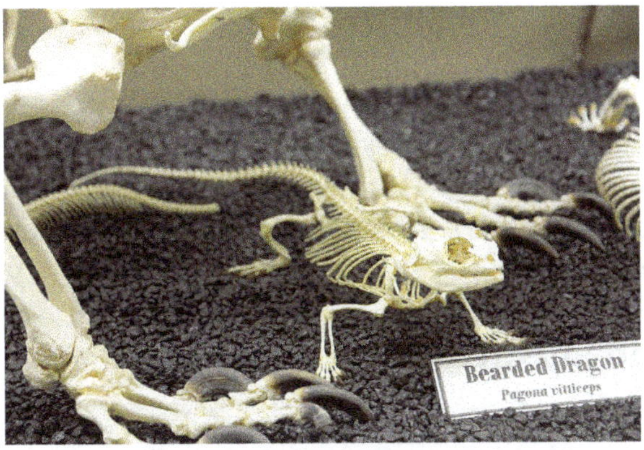

Skeleton of bearded dragon (pogona sp.) on display at the Museum of Osteology.

Some lizards, particularly iguanas, have retained a photosensory organ on the top of their heads called the parietal eye, a basal ("primitive") feature also present in the tuatara. This "eye" has only a rudimentary retina and lens and cannot form images, but is sensitive to changes in light and dark and can detect movement. This helps them detect predators stalking it from above.

Venom

Some lizards including the gila monster are venomous.

Until 2006 it was thought that among lizards, only the Gila monster and the Mexican beaded lizard were venomous. However, several species of monitor lizards, including the Komodo dragon, produce powerful venom in their oral glands. Lace monitor venom, for instance, causes swift loss of consciousness and extensive bleeding through its pharmacological effects, both lowering blood pressure and preventing blood clotting. Nine classes of toxin known from snakes are produced by lizards. The range of actions provides the potential for new medicinal drugs based on lizard venom proteins.

Genes associated with venom toxins have been found in the salivary glands on a wide range of lizards, including species traditionally thought of as non-venomous, such as iguanas and bearded dragons. This suggests that these genes evolved in the common ancestor of lizards and snakes, some 200 million years ago (forming a single clade, the Toxicofera). However, most of these putative venom genes were "housekeeping genes" found in all cells and tissues, including skin and cloacal scent glands. The genes in question may thus be evolutionary precursors of venom genes.

Respiration

Recent studies on the lung anatomy of the savannah monitor and green iguana found them to have a unidirectional airflow system, which involves the air moving in a loop through the lungs when breathing. This was previously thought to only exist in the archosaurs (crocodilians and birds). This may be evidence that unidirectional airflow is an ancestral trait in diapsids.

Reproduction and Lifecycle

Trachylepis maculilabris skinks mating.

As with all amniotes, lizards rely on internal fertilisation and copulation involves the male inserting one of his hemipenes into the female's cloaca. The majority of species are oviparous (egg laying). The female deposits the eggs in a protective structure like a nest or crevice or simply on the ground. Depending on the species, clutch size can vary from 4–5 percent of the females body weight to 40–50 percent and clutches range from one or a few large eggs to dozens of small ones.

Two pictures taken on an eastern fence lizard egg

In most lizards, the eggs have leathery shells to allow for the exchange of water, although more arid-living species have calcified shells to retain water. Inside the eggs, the embryos use nutrients from the yolk. Parental care is uncommon and the female usually abandons the eggs after laying them. Brooding and protection of eggs does occur in some species. The female prairie skink uses respiratory water loss to maintain the humidity of the eggs which facilitates embryonic development. In lace monitors, the young hatch close to 300 days, and the female returns to help them escape the termite mound were the eggs were laid.

Around 20 percent of lizard species reproduce via viviparity (live birth). This is particularly common in Anguimorphs. Viviparous species give birth to relatively developed young which look like miniature adults. Embryos are nourished via a placenta-like structure. A minority of lizards have parthenogenesis (reproduction from unfertilised eggs). These species consist of all females who reproduce asexually with no need for males. This is known in occur in various species of whiptail lizards. Parthenogenesis was also recorded in species that normally reproduce sexually. A captive female Komodo dragon produced a clutch of eggs, despite being separated from males for over two years.

Sex determination in lizards can be temperature-dependent. The temperature of the eggs' micro-environment can determine the sex of the hatched young: low temperature incubation produces more females while higher temperatures produce more males. However, some lizards have sex chromosomes and both male heterogamety (XY and XXY) and female heterogamety (ZW) occur.

Behaviour

Diurnality and Thermoregulation

The majority of lizard species are active during the day, though some are active at night, notably geckos. As ectotherms, lizards have a limited ability to regulate their body temperature, and must seek out and bask in sunlight to gain enough heat to become fully active.

Territoriality

Fighting male sand lizards

Most social interactions among lizards are between breeding individuals. Territoriality is common and is correlated with species that use sit-and-wait hunting strategies. Males establish and maintain territories that contain resources which attract females and which they defend from other males. Important resources include basking, feeding, and nesting sites as well as refuges from predators. The habitat of a species affects the structure of territories, for example, rock lizards have territories atop rocky outcrops. Some species may aggregate in groups, enhancing vigilance and lessening the risk of predation for individuals, particularly for juveniles. Agonistic behaviour typically occurs between sexually mature males over territory or mates and may involve displays, posturing, chasing, grappling and biting.

A Green Anole (*Anolis carolinensis*) signalling with its extended dewlap

Communication

Lizards signal both to attract mates and to intimidate rivals. Visual displays include body postures and inflation, push-ups, bright colours, mouth gapings and tail waggings. Male anoles and iguanas have dewlaps or skin flaps which come in various sizes, colours and patterns and the expansion of the dewlap as well as head-bobs and body movements add to the visual signals. Some species have deep blue dewlaps and communicate with ultraviolet signals. Blue-tongued skinks will flash their tongues as a threat display. Chameleons are known to change their complex colour patterns when communicating, particularly during agonistic encounters. They tend to show brighter colours when displaying aggression and darker colours when they submit or "give up".

Several gecko species are brightly coloured; some species tilt their bodies to display their coloration. In certain species, brightly coloured males turn dull when not in the presence of rivals or females. While it is usually males that display, in some species females also use such communication. In the bronze anole, head-bobs are a common form of communication among females, the speed and frequency varying with age and territorial status. Chemical cues or pheromones are also important in communication. Males typically direct signals at rivals, while females direct them at potential mates. Lizards may be able to recognise individuals of the same species by their scent.

Acoustic communication is less common in lizards. Hissing, a typical reptilian sound, is mostly produced by larger species as part of a threat display, accompanying gaping jaws. Some groups, particularly geckos, snake-lizards, and some iguanids, can produce more complex sounds and vocal apparatuses have independently evolved in different groups. These sounds are used for courtship, territorial defense and in distress, and include clicks, squeaks, barks and growls. The mating call of the male tokay gecko is heard as "tokay-tokay!". Tactile communication involves individuals rubbing against each other, either in courtship or in aggression. Some chameleon species communicate with one another by vibrating the substrate that they are standing on, such as a tree branch or leaf.

Ecology

Lizard in tree. Many species are tree-dwelling.

Distribution and Habitat

Lizards are found worldwide, excluding the far north and Antarctica, and some islands. They can be found in elevations from sea level to 5,000 m (16,000 ft). They prefer warmer, tropical climates but are adaptable and can live in all but the most extreme environments. Lizards also exploit a number of habitats; most primarily live on the ground, but others may live in rocks, on trees, underground and even in water. The marine iguana is adapted for life in the sea.

Diet

Western green lizard ambushes its grasshopper prey.

The majority of lizard species are predatory and the most common prey items are small, terrestrial invertebrates, particularly insects. Many species are sit-and-wait predators though others may be more active foragers. Chameleons prey on numerous insect species, such as beetles, grasshoppers and winged termites as well as spiders. They rely on persistence and ambush to capture these prey. An individual perches on a branch and stays perfectly still, with only its eyes moving. When an insect lands, the chameleon focuses its eyes on the target and slowly moves towards it before projecting its long sticky tongue which, when hauled back, brings the attach prey with it. Geckos feed on crickets, beetles, termites and moths.

Termites are an important part of the diets of some species of Autarchoglossa, since, as social insects, they can be found in large numbers in one spot. Ants may form a prominent part of the diet of some lizards, particularly among the lacertas. Horned lizards are also well known for specializing on ants. Due to their small size and indigestible chitin, ants must be consumed in large amounts, and ant-eating lizards have larger stomachs than even herbivorous ones. Species of skink and alligator lizards eat snails and their power jaws and molar-like teeth are adapted for breaking the shells.

Young Komodo dragon feeding on a water buffalo carcass.

Larger species, such as monitor lizards, can feed on larger prey including fish, frogs, birds, mammals and other reptiles. Prey may be swallowed whole and torn into smaller pieces. Both bird and reptile eggs may also be consumed as well. Gila monsters and beaded lizards climb trees to reach both the eggs and young of birds. Despite being venomous, these species rely on their strong jaws to kill prey. Mammalian prey typically consists of rodents and leporids; the Komodo dragon can kill prey as large as water buffalo. Dragons are prolific scavengers, and a single decaying carcass can attract several from 2 km (1.2 mi) away. A 50 kg (110 lb) dragon is capable of consuming a 31 kg (68 lb) carcass in 17 minutes.

Marine iguana foraging under water at Galápagos Islands, Ecuador.

Around 2 percent of lizard species, including many iguanids, are herbivores. Adults of these species eat plant parts like flowers, leaves, stems and fruit, while juveniles eat more insects. Plant parts can be hard to digest and as they get closer to adulthood, juvenile iguanas eat faeces from adults to acquire the microflora necessary for their transition to a plant-based diet. Perhaps the most herbivorous species is the marine iguana which dives 15 m (49 ft) to forage for algae, kelp and other marine plants. Some non-herbivorous species supplement their insect diet with fruit, which is easily digested.

Antipredator Adaptations

The frilled-neck lizard with fully extended frill: The frilled serves to make it look bigger than it actually is.

Lizards have a variety of antipredator adaptations, including running and climbing, venom, camouflage, tail autotomy, and reflex bleeding.

Camouflage

The flat-tail horned lizard's body is flattened and fringed to minimise its shadow.

Lizards exploit a variety of different camouflage methods. Many lizards are disruptively patterned. In some species, such as Aegean wall lizards, individuals vary in colour, and select rocks which best match their own colour to minimise the risk of being detected by predators. The Moorish gecko is able to change colour for camouflage: when a light-coloured gecko is placed on a dark surface, it darkens within an hour to match the environment. The chameleons in general use their ability to change their coloration for signalling rather than camouflage, but some species such as Smith's dwarf chameleon do use active colour change for camouflage purposes.

The flat-tail horned lizard's body is coloured like its desert background, and is flattened and fringed with white scales to minimise its shadow.

Autotomy

Many lizards, including geckos and skinks, are capable of shedding their tails (autotomy). The detached tail, sometimes brilliantly coloured, continues to writhe after detaching, distracting the predator's attention from the fleeing prey. Lizards partially regenerate their tails over a period of weeks. Some 326 genes are involved in regenerating lizard tails. The fish-scale gecko *Geckolepis megalepis* sheds patches of skin and scales if grabbed.

Escape, Playing Dead, Reflex Bleeding

Many lizards attempt to escape from danger by running to a place of safety;[b] for example, wall lizards can run up walls and hide in holes or cracks. Horned lizards adopt differing defences for specific predators. They may play dead to deceive a predator that has caught them; attempt to outrun the rattlesnake, which does not pursue prey; but stay still, relying on their cryptic coloration, for *Masticophis* whip snakes which can catch even swift prey. If caught, some species such as the greater short-horned lizard puff themselves up, making their bodies hard for a narrow-mouthed predator like a whip snake to swallow. Finally, horned lizards can squirt blood at cat and dog predators from a pouch beneath its eyes, to a distance of about two metres (6.6 feet); the blood tastes foul to these attackers.

Crocodile

The American alligator (Alligator mississippiensis) is found in the southeastern United States.

Crocodile, (order Crocodylia, or Crocodilia) is any of 23 species of generally large, ponderous, amphibious animals of lizard-like appearance and carnivorous habit belonging to the reptile order Crocodylia. Crocodiles have powerful jaws with many conical teeth and short legs with clawed webbed toes. They share a unique body form that allows the eyes, ears, and nostrils to be above the water surface while most of the animal is hidden below. The tail is long and massive, and the skin is thick and plated.

Crocodiles are a living link with the dinosaur-like reptiles of prehistoric times and are the nearest living relatives of the birds. A large variety of crocodilian fossils have been discovered that date back 200 million years to the Late Triassic Epoch. Fossil evidence also suggests that three major radiations occurred. Only one of the four suborders of crocodiles has survived to modern times. The order Crocodylia includes the "true crocodiles," alligators, caimans, and gavials.

General Features

Size Range and Diversity of Structure

Crocodiles are the largest and the heaviest of present-day reptiles. The largest representatives, the Nile crocodile (Crocodylus niloticus) of Africa and the estuarine (or saltwater) crocodile (C. porosus) of Australia, attain lengths of up to 6 metres (20 feet) and weigh over 1,000 kg (about 2,200 pounds). Some fossil forms (such as Deinosuchus and Sarcosuchus) may have been between 10 and 12 metres (33 and 40 feet) long. In comparison, the smallest species, the smooth-fronted caiman (Paleosuchus) and the dwarf crocodile (Osteolaemus tetraspis), reach about 1.7 metres (about 6 feet) in length as adults.

Nile crocodile (Crocodylus niloticus).

All crocodiles have a relatively long snout, or muzzle, which varies considerably in shape and proportion. The scales that cover most of the body generally are arranged in a regular pattern, and thick, bony plates occur on the back. Families and genera are primarily distinguished by differences in skull anatomy. Species are identified principally by the proportions of the snout; by the bony structures on the dorsal, or upper, surface of the snout; and by the number and the arrangement of the scales.

The Nile crocodile (Crocodylus niloticus) inhabits a wide swath of eastern and sub-Saharan Africa and Madagascar. It is one of the largest crocodiles, growing to lengths up to 6 metres (20 feet) and a mass of more than 1,000 kg (about 2,200 pounds).

Crocodile

Distribution and Abundance

Crocodiles are found mainly in the lowland, humid tropics of the Northern and Southern hemispheres. The "true crocodiles" (family Crocodylidae) occur in most of Africa south of the Sahara, Madagascar, India, Sri Lanka, Southeast Asia, the East Indies, northern Australia, Mexico and Central America, the West Indies, and northern South America. In the family Alligatoridae, most caimans are confined to the tropical areas of Central and South America, though the ranges of the broad-snouted caiman (Caiman latirostris) and Jacaré caiman (C. yacare) extend into temperate areas of South America. The American alligator (Alligator mississippiensis) and the Chinese alligator (A. sinensis) also occur in temperate regions. In the family Gavialidae, the Indian gavial (Gavialis gangeticus) is found in Pakistan, northern India, Nepal, Bhutan, Bangladesh, and Myanmar.

Gharial, or gavial (Gavialis gangeticus).

Throughout their range, crocodile populations have declined as human occupation and land use change has reduced their habitat. Many crocodilian species have been greatly depleted by overhunting for their valuable skins—which provide leather for handbags, shoes, belts, and other articles. Local use of crocodiles for meat and medicines is also widespread. People living near crocodiles often dislike them because they become entangled in fishnets, prey on pets and livestock, and occasionally kill people.

Sustainable harvesting, regulated trade, and education have become valuable components of crocodilian conservation. Worldwide, various programs exist to provide incentives and economic benefits to communities that conserve habitat. For example, some communities in the Philippines are paid for each egg made safe from collectors or those intent on destroying nests. Crocodilian conservation has become a model for sustainable resource use. Commercial production by captive breeding, the collection of excess eggs from the wild, and regulated hunting yield 800,000–1,000,000 legal skins each year to international markets.

Life Cycle

The young crocodile emerges from the egg with a length of 20–30 cm (8–12 inches). At first it remains with its siblings concealed at the edge of its water habitat in order to avoid various predators. Principal among these are fishes and birds, but larger crocodiles also prey upon the young. Cannibalism and social exclusion are thought to greatly affect population dynamics and regulate population growth. During the first three to four years, the young increase in length by about 30 cm (about 1 foot) per year. The growth rate then gradually decreases, but growth can continue throughout life. Sexual maturity occurs at about age 10 and at a body length of about 1.5–3 metres (5–10 feet).

Crocodiles in captivity are known to have reached ages of 70 or more years, but longevity in the wild is poorly known. Life span estimates in the wild are based on growth rates, and limited studies of bone growth rings suggest that the life spans of wild crocodiles and those in captivity may be similar. A Nile crocodile (Crocodylus niloticus) or an estuarine (or saltwater) crocodile (C. porosus) 6 metres (about 20 feet) long may live as long as 80 years. On average, the life span of alligators and caimans is thought to be between 30 and about 60 years, whereas the true crocodiles are thought to live 50 to 75 years.

Behaviour

Crocodiles are mostly nocturnal animals. They are predators and spend most of their time in the water; although they are also known to make journeys of several kilometres over land. In the first weeks of life, crocodiles eat insects, crustaceans, snails, small fishes, frogs, and tadpoles. Older crocodiles mainly eat fish and are more apt to prey upon waterfowl and on mammals. Occasionally, a member of one of the larger species eats a human, though such incidents happen so infrequently that crocodiles cannot be generally regarded as man-eaters.

Crocodiles capture water animals in their jaws with a sideways movement of the muzzle. They have sensitive pressure receptors located in pits in the scales around the mouth that detect motion; these structures assist in the capture of prey in dark or murky water. To catch land animals, a crocodile floats passively or remains motionless at the edge of the water where prey habitually drink. With a sudden lunge, it seizes an unsuspecting animal and drowns it. If the prey is large, a crocodile may grip portions of the victim in its jaws and rotate rapidly in the water to tear the prey apart.

Nile crocodile (Crocodylus niloticus) swallowing a fish.

Crocodiles thermoregulate by alternately sunning themselves and retiring to shaded areas or cooler water. Larger individuals maintain stable body temperatures in the preferred range of 30–32 °C (86–90 °F) for several hours, even overnight. As a result, these individuals enjoy increased metabolic efficiency. Several studies of wild populations suggest that complex social relations between individuals are expressed as dominance hierarchies that allow dominant animals better access to preferred sunning and nesting sites. Some crocodiles also dig burrows into the banks of lakes or rivers. Burrows may extend for several metres in length and end in a chamber where individuals seek refuge from drought or cold.

Many crocodiles vocalize to communicate. The young of various species use several squeaking and grunting sounds, and adults may grunt, growl, and hiss. For example, Siamese crocodiles and caimans emit a loud hiss when threatened, and hatchlings of most species make sounds described as grunts or quacks. In addition, members of both sexes may produce a loud roar during the breeding season. A roaring crocodile tenses the muscles of its body so that the head and tail rise high out of the water. The flanks may vibrate so violently that water is sprayed high into the air from each side. Sounds, including roars, may be provoked by any loud noise. Many species will respond to gunshots, motors, and even people mimicking crocodile sounds. These animals also appear to communicate using chemical signals. Glands in the mandible and cloaca excrete oily chemicals that have a poorly understood function in communication.

The crocodilian brain is relatively tiny compared with the size of the body; the brain of an alligator that is 4 metres (13 feet) long weighs only 11 grams (0.02 pound). Nevertheless, crocodiles are capable of complex behaviours. They are often curious and show evidence of rapid learning. Captive individuals of some species are known to recognize their keepers and show neither fear nor aggressiveness. These animals beg for food, and some even permit themselves to be petted.

Locomotion

Crocodilian locomotion is the product of a unique body form. The lateral processes of the spine are joined to interlocking bony plates of the dorsal scales by complex ligaments. This arrangement produces a semirigid “I-beam” structure that encloses the back muscles. This structure is rigid but flexible, and it allows an efficient transfer of energy from the tail when swimming and an erect body posture when walking. When swimming, the crocodile places its legs back against the sides of the body and moves forward by means of lateral wavelike motions of the tail. When walking on land, crocodiles hold themselves high on all four legs. The characteristic sinusoidal (side-to-side) flexure of the body is caused by the movement of a front leg in concert with the opposing hind leg during each step. A cantilevered tail also balances the body. When moving quickly into the water from a bank, crocodiles slide on their bellies and push themselves forward with the feet. Crocodiles are also capable of galloping short distances.

Reproduction

Crocodiles are sexually dimorphic, and adult males are larger than females. Copulation occurs in the water, and it is preceded by a complex courtship in which the animals signal each other using changes in body profile, touch, and vocalization. Chemical signals are probably also part of courtship. The male then mounts the back of the female, and both animals rotate their tails so that the respective cloacae are brought into contact and intromission of the male erectile organ is achieved.

All crocodiles lay hard-shelled eggs, which may weigh 50–160 grams (0.1–0.4 pound) each. A female lays an average of 12–48 eggs per nest, depending upon her age, size, and species. Two general forms of nest building are known. Some species, such as the Nile crocodile (C. niloticus), dig a hole in the ground and refill it with dirt after the eggs are deposited. Others, such as the estuarine (or saltwater) crocodile (C. porosus), build a mound of plant material and soil. The eggs are deposited into the mound, and the sun's heat, the warm environment, and the natural decay of vegetation maintain a warm temperature that aids the development of the embryo. Egg incubation takes from 55 to 100 days. As is the case for many other reptiles, the sex of the developing embryo is determined by the temperature of the incubating egg at a critical time, which is during the first half of its incubation period. Cooler and warmer temperatures produce females, intermediate temperatures produce males, and temperatures near 31 °C (88 °F) produce both sexes. Until hatching occurs, the female usually remains close to the nest to protect the eggs from predators.

After two or three months, the young are fully developed and ready to hatch. While still in the egg, the young may utter squeaks, perhaps signaling that they are ready to emerge. The adult female removes the dirt or other debris from the eggs and assists the hatchlings to the water. In many cases, the female opens the eggs carefully with her tongue and carries the hatchlings in her mouth. The female remains close to her offspring and provides protection from predators for several weeks to months. As the young grow, they disperse, but they may remain near their nest site for over a year. In captivity, males have also been observed assisting in the hatching process and providing hatchling care.

Ecology

Crocodiles are inhabitants of swamps, lakes, and rivers, although some species make their way to brackish water or to the sea. The estuarine (or saltwater) crocodile (C. porosus) and the American crocodile (C. acutus) are capable of living in marine waters and may swim miles out to sea, although both species normally occupy brackish and freshwater habitats. Glands in the tongue allow the excretion of excess salt. The smooth-fronted caiman (Paleosuchus trigonatus) of South America prefers rocky, fast-flowing rivers. In West Africa the dwarf crocodile (Osteolaemus tetraspis) is found principally in the rivers of the forest regions. Crocodiles may affect nutrient cycling, ecosystem function, and fisheries productivity in the areas they inhabit.

American crocodile (Crocodylus acutus).

Form and Function

The crocodilian form is adapted to an amphibious way of life. The body is elongated, and its long, muscular tail is well suited to rapid swimming.

Nile crocodile (Crocodylus niloticus).

The external nostril openings, the eyes, and the ear openings are the highest parts of the head. These important sense organs remain above the water surface even when the rest of the head and body are submerged. The two nostril openings are close together on a raised portion at the point of the muzzle. When the animal dives, these openings may be closed by membranous flaps to keep water out. A long bone-enclosed nasal passage leads from the exterior nostril openings to the interior nostril openings, or choanae, located at the extreme posterior end of the palate; a membranous flap in front of the choanae constitutes the posterior closure of the mouth cavity. As a result, the crocodile can breathe even if its mouth is open underwater.

Like many nocturnal animals, crocodiles have eyes with vertical, slit-shaped pupils; these narrow in bright light and widen in darkness, thus controlling the amount of light that enters. On the back wall of the eye, the tapetum lucidum reflects incoming light, thus utilizing the small amount of light available at night to best advantage. In addition to the protection provided by the upper and lower eyelids, the nictitating membrane (that is, a thin, translucent eyelid) may be drawn over the eye from the inner corner while the lids are open. The delicate eyeball surface is thus protected under the water, while a certain degree of vision is still possible. Unlike the ears of other modern reptiles, those of the crocodile have a movable, external membranous flap that protects the ears from the water. The sense of smell is well developed and may even operate in the egg prior to hatching.

The outer margin of the jaws in most species is irregular. Each jaw carries a row of conical teeth, which may number more than 100 in species with very long muzzles. The teeth are held in sockets and replaced continuously; new teeth grow from below and force the older ones out. The thick, fleshy tongue is firmly attached to the floor of the mouth, and it is nearly immobile.

Crocodile

The posterior portion of the head forms a flat plate that is attached to a short, powerful neck. On the dorsal side of the neck are two groups of bony scales called scutes. Small postoccipital scutes are located just behind the head and are present in all crocodiles except the estuarine crocodile. Behind the postoccipital scutes are the larger nuchal scutes, which in some species are connected to the adjacent horny plates of the back.

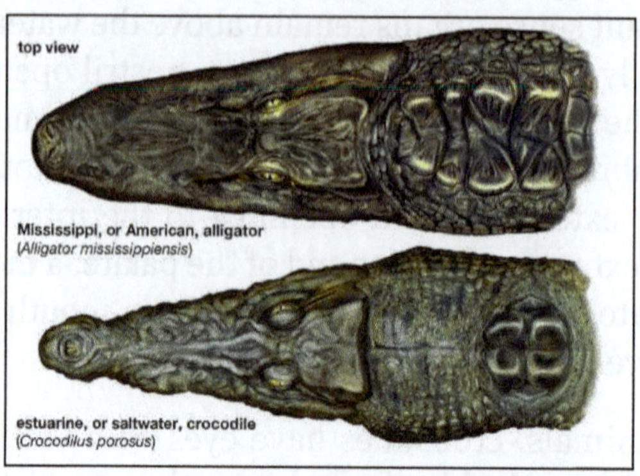

A view from above shows the different snouts of an alligator and a crocodile.

The upper surfaces of the back and tail are covered with large, rectangular horny plates arranged regularly in longitudinal and transverse rows. Most of the dorsal plates have a longitudinal ridge, or keel. Under these plates lie bony structures called osteoderms of about the same size. This configuration occurs in all but one species; in the estuarine crocodile, the bony plates are smaller.

The entire underside of the crocodile has a regular pattern of scales, which are smaller than those on the upper surface. These scales are rectangular, entirely smooth, and contain little or no bone material. An exception to this condition occurs in caimans of the genera Melanosuchus, Caiman, and Paleosuchus, in which the surface plates on the lower side are also bony. The cloaca—a chamber containing the genital, anal, and urogenital openings—extends longitudinally within the body; it is surrounded by an oval area of small scales on the underside of the body. The cloacal vent is slightly posterior to the attachment of the hind legs at the base of the tail. In contrast to the back and belly, the sides of the body have mostly small knobby scales. As a result, the flanks are distensible, which is necessary for breathing and for the expansion of the body that occurs in gravid females.

The legs of the crocodile are short but powerful. The forefeet have five toes—the usual number for class Reptilia. The hind legs are more powerfully developed than the front pair, and the hind feet have only four toes, which are wholly or partially webbed. The upper part of the tail carries two crests of high triangular scales that gradually merge near the middle of the tail.

Structurally, the heart of the crocodile is markedly different from that of other reptiles. In all other reptiles, oxygenated blood mixes with deoxygenated blood in the ventricles, because the ventricles are only partially divided from each other. In crocodiles, both atria and both ventricles are completely separate. Nevertheless, a connection exists between arterial and venous circulation by way of the foramen of Panizza, which opens between the two vessels leading separately from the ventricles. This connection operates when the crocodile holds its breath. It allows blood to bypass the lungs when the animal is submerged, and this structure has the effect of stabilizing blood oxygen levels.

Turtle

Turtle is any aquatic or terrestrial reptile of the order Testudines (or Chelonia), characterized by toothless jaws with horny beaks and generally having a body shielded by a special bony or cartilagenous shell. Tortoise and terrapin are the names for two subgroups commonly recognized within Testudines. Tortoise is the common name for any land-dwelling turtle, especially those belonging to the family Testudinidae. Terrapin is the common name for large freshwater or brackish water turtles belonging to the family Emydidae, especially the genus Malaclemys, and sometimes the genus Pseudemys (or Chrysemys).

As they advance their own survival and reproduction, turtles also play a vital role in food chains, both as herbivores and carnivores and as prey (particularly as vulnerable hatchlings). Their unique adaptations also provide unique aesthetic and practical values to human beings, with their shells collected as ornaments, and their behaviors (such as the new hatchlings making their way on the beach to the ocean) adding to human fascination with nature. Turtles have historically served as food or skinned for leather.

Not all turtles (also known technically as chelonians) have armor-like shells. The Trionychidae family has members commonly referred to as "softshell turtles," such as with the North American genus Apalone, because their carapace (outer, upper covering) lacks scutes (scales). The Australasian pig-nose turtle, Carettochelys insculpta, found in New Guinea and Australia and also known as the "plateless turtle," is a species of soft-shelled turtle whose gray carapace has a leathery texture. The leatherback sea turtle (Dermochelys coriacea), found in all tropical and subtropical oceans, has a shell that lacks the bony scutes of other turtles, comprising mainly of connective tissue.

Turtles are found in most parts of the world, and there are about 300 species alive today. Turtles are ectothermic or cold-blooded, which means that their body temperature changes with their environment. They all breathe with lungs and, whether aquatic or terrestrial, reproduction involves laying eggs on land, including the highly aquatic sea turtles.

While turtles offer important ecological, commercial, and aesthetic values, and as a group have survived for millions of years, today many of the species are rare or endangered. This is largely due to anthropogenic factors, such as loss of habitat, pollution, and accidental catch in commercial fishing.

Anatomy and Morphology

As reptiles, turtles are tetrapods (four-legged vertebrates) and amniotes (animals whose embryos are surrounded by an amniotic membrane that encases it in amniotic fluid.

Turtles vary widely in size, although marine turtles tend to be relatively bigger animals than their land and freshwater relatives.

The largest extant turtle is a marine turtle, the great leatherback sea turtle, which reaches a shell length of over 2.7 meters (8.8 feet) and weight of 900 kilograms (kg) (2,000 lb)—the world's fourth largest reptile, behind the larger crocodiles. Freshwater turtles are generally smaller, but with the largest species, the Asian softshell turtle Pelochelys cantorii, a few individuals have been reported to measure up to 200 centimeters (80 inches). This dwarfs even the better-known alligator snapping turtle, the largest chelonian in North America, which attains a shell length of up to 80 centimeters (31½ inches) and a weight of about 60 kg (170 lb).

Giant tortoises of the genera Geochelone, Meiolania, and others were relatively widely distributed around the world into prehistoric times, and are known to have existed in North and South America, Australia, and Africa. They became extinct at the same time as the appearance of humans, and it is assumed that humans hunted them for food. The only surviving giant tortoises are on the Seychelles and Galápagos Islands and can grow to over 130 centimeters (50 inches) in length, and weigh about 300 kg (670 lb).

The largest known chelonian in the fossil record was *Archelon ischyros,* a Late Cretaceous sea turtle known to have been up to 4.6 meters (15 feet) long.

The smallest turtle is the speckled padloper tortoise of South Africa. It measures no more than 8 centimeters (3 inches) in length and weighs about 140 grams (5 ounces). Two other species of small turtles are the American mud turtles and musk turtles that live in an area that ranges from Canada to South America. The shell length of many species in this group is less than 13 centimeters (5 inches) in length.

Neck Folding

Turtles are broken down into two groups, according to how they evolved a solution to the problem of withdrawing their neck into their shell. In most turtles, the Cryptodira, the neck is folding under their spine, being drawn directly back into the shell in an S-shaped curve. In the remainder, the Pleurodira, or side-necked turtles, the neck is tucked next to the shoulder. Ancestral turtles are believed to have not been able to retract their neck.

Closeup head-on view of a common snapping turtle (Chelydra serpentina), taken near the St. Lawrence River in northern New York State

Head

Most turtles that spend most of their life on land have their eyes looking down at objects in front of them. Some aquatic turtles, such as snapping turtles and soft-shelled turtles, have eyes closer to the top of the head. These species of turtles can hide from predators in shallow water where they lie entirely submerged except for their eyes and nostrils.

Sea turtles possess glands near their eyes that produce salty tears that rid their body of excess salt taken in from the water they drink.

Turtles are thought to have exceptional night vision due to the unusually large amount of rod cells in their retinas. Normal daytime vision is marginal at best due to their color-blindness and poor visual acuity. In addition to daytime vision problems, turtles have very poor pursuit movement abilities, which are normally reserved for predators that hunt quick moving prey. However, carnivorous turtles can move their heads quickly to snap.

Turtles have a rigid, toothless beak. Turtles use their jaws to cut and chew food. Instead of teeth, the upper and lower jaws of the turtle are covered by horny ridges. Carnivorous turtles usually have knife-sharp ridges for slicing through their prey. Herbivorous turtles have serrated edged ridges that help them cut through tough plants. Turtles use their tongues to swallow food, but, unlike most reptiles, they cannot extend their tongues to catch food.

Shell

The upper shell or upper outer covering of a turtle is called the carapace. The lower shell that encases the belly is called the plastron. The carapace and plastron are joined together on the turtle's sides by bony structures called bridges.

The inner layer of a turtle's shell is made up of about 60 bones that include portions of the backbone and the ribs, meaning the turtle cannot crawl out of its shell.

In most turtles, the outer layer of the shell is covered by horny scales called scutes that are part of its outer skin, or epidermis. Scutes are made up of a fibrous protein called keratin that also makes up the scales of other reptiles. These scutes overlap the seams between the shell bones and add strength to the shell. Some turtles do not have horny scutes. For example, the leatherback sea turtle and the soft-shelled turtles have shells covered with leathery skin instead.

The shape of the shell gives helpful clues to how the turtle lives. Most tortoises have a large dome-shaped shell that makes it difficult for predators to crush the shell between their jaws. One of the few exceptions is the African pancake tortoise, which has a flat, flexible shell that allows it to hide in rock crevices. Most aquatic turtles have flat, streamlined shells, which aid in swimming and diving. American snapping turtles and musk turtles have small, cross-shaped plastrons that give them more efficient leg movement for walking along the bottom of ponds and streams.

The color of a turtle's shell may vary. Shells are commonly colored brown, black, or olive green. In some species, shells may have red, orange, yellow, or gray markings and these markings are often spots, lines, or irregular blotches. One of the most colorful turtles is the eastern painted turtle, which includes a yellow plastron and a black or olive shell with red markings around the rim.

Tortoises, being land-based, have rather heavy shells. In contrast, aquatic and soft-shelled turtles have lighter shells that help them avoid sinking in water and allow them to swim faster with more agility. These lighter shells have large spaces called fontanelles between the shell bones. The shell of a leatherback turtle is extremely light because they lack scutes and contain many fontanelles.

Skin and Molting

The outer layer of the shell is part of the skin. Every scute (or plate) on the shell corresponds to a single modified scale. The remainder of the skin is composed of skin with much smaller scales, similar to the skin of other reptiles. Growth requires molting of a turtle's skin, although not the scutes.

Turtles, including terrapins, do not molt their skins all in one event, as snakes do, but continuously, in small pieces. When kept in aquaria, small sheets of dead skin can be seen in the water when it has been sloughed off, (often appearing to be a thin piece of plastic), and often when the animal deliberately rubs itself against a piece of wood or stone. Tortoises also shed skin, but a lot of dead skin is allowed to accumulate into thick knobs and plates that provide protection to parts of the body outside the shell.

The scutes on the shell are never molted, and, as they accumulate over time, the shell becomes thicker. By counting the rings formed by the stack of smaller, older scutes on top of the larger, newer ones, it is possible to estimate the age of a turtle, if you know how many scutes are produced in a year. This method is not very accurate, partly because growth rate is not constant, but also because some of the scutes eventually fall away from the shell.

Limbs

Terrestrial tortoises have short, sturdy feet. Tortoises are famous for moving slowly. In part this is because of their heavy, cumbersome shell. However, it also is a result of the relatively inefficient sprawling gait that they have, with the legs being bent, as with lizards rather than being straight and directly under the body, as is the case with mammals.

The amphibious turtles normally have limbs similar to those of tortoises except that the feet are webbed and often have long claws. These turtles swim using all four feet in a way similar to the dog paddle, with the feet on the left and right side of the body alternately providing thrust. Large turtles tend to swim less than smaller ones, and the very big species, such as alligator snapping turtles, hardly swim at all, preferring to simply walk along the bottom of the river or lake. As well as webbed feet, turtles also have very long claws, used to help them clamber onto riverbanks and floating logs, upon which they like to bask. Male turtles tend to have particularly long claws, and these appear to be used to stimulate the female while mating. While most turtles have webbed feet, a few turtles, such as the pig-nose turtles, have true flippers, with the digits being fused into paddles and the claws being relatively small. These species swim in the same way as sea turtles.

Sea turtles are almost entirely aquatic and instead of feet they have flippers. Sea turtles "fly" through the water, using the up-and-down motion of the front flippers to generate thrust; the back feet are not used for propulsion but may be used as rudders for steering. Compared with freshwater turtles, sea turtles have very limited mobility on land, and apart from the dash from the nest to the sea as hatchlings, male sea turtles normally never leave the sea. Females must come back onto land to lay eggs. They move very slowly and laboriously, dragging themselves forwards with their flippers. The back flippers are used to dig the burrow and then fill it back with sand once the eggs have been deposited.

Ecology and Life History

Even though many spend large amounts of their lives underwater, all turtles are air-breathing reptiles, and must surface at regular intervals to refill their lungs with fresh air. They can also spend a lot of their lives on dry land.

An American map turtle hatchling.

Some species of Australian freshwater turtles have large cloacal cavities that are lined with many finger-like projections. These projections, called "papillae," have a rich blood supply, and serve to increase the surface area of the cloaca. The turtles can take up dissolved oxygen from the water using these papillae, in much the same way that fish use gills to respire.

Turtles lay eggs, like other reptiles, which are slightly soft and leathery. The eggs of the largest species are spherical, while the eggs of the rest are elongated. Their albumen is white and contains a different protein than do bird eggs, such that it will not coagulate when cooked. Turtle eggs prepared to eat consist mainly of yolk.

In some species, temperature determines whether an egg develops into a male or a female: a higher temperature causes a female, a lower temperature causes a male.

Turtles lay the eggs on land. Large numbers of eggs are deposited in holes dug into mud or sand. They are then covered and left to incubate by themselves. When the turtles hatch they squirm their way to the surface and make for the water. There are no known species wherein the mother cares for the young.

Sea turtles lay their eggs on dry sandy beaches, and are highly endangered largely as a result of beach development and overhunting. Turtles can take many years to reach breeding age. Often turtles only breed every few years or more.

Researchers have recently discovered a turtle's organs do not gradually break down or become less efficient over time, unlike most other animals. It was found that the liver, lungs, and kidneys of a centenarian turtle are virtually indistinguishable from those of its immature counterpart. This has inspired genetic researchers to begin examining the turtle genome for genes related to longevity.

Batrachology

Batrachology is the branch of zoology worried about the study of creatures of land and water including frogs and amphibians, salamanders, newts, and caecilians. It is a sub-teach of herpetology, which incorporates additionally non-avian reptiles like snakes, reptiles, amphisbaenids, turtles, reptiles, tortoises, crocodilians, and the tuatara. Batrachologists may examine the advancement, nature, ethology, or life systems of creatures of land and water.

Creatures of land and water are cold-blooded vertebrates to a great extent found in soggy natural surroundings albeit numerous species have uncommon social adjustments that enable them to live in deserts, trees, underground and in districts with wide occasional varieties in temperature. There are more than 7250 types of creatures of land and water.

Amphibian

Amphibian is any member of the group of vertebrate animals characterized by their ability to exploit both aquatic and terrestrial habitats.

Green frog (Rana clamitans melanota).

More than 6,500 species of living amphibians are known. First appearing about 340 million years ago during the Middle Mississippian Epoch, they were one of the earliest groups to diverge from ancestral fish-tetrapod stock during the evolution of animals from strictly aquatic forms to terrestrial types. Today amphibians are represented by frogs and toads (order Anura), newts and salamanders (order Caudata), and caecilians (order Gymnophiona). These three orders of living amphibians are thought to derive from a single radiation of ancient amphibians, and although strikingly different in body form, they are probably the closest relatives to one another. As a group, the three orders make up subclass Lissamphibia. Neither the lissamphibians nor any of the extinct groups of amphibians were the ancestors of the group of tetrapods that gave rise to reptiles. Though some aspects of the biology and anatomy of the various amphibian groups might demonstrate features possessed by reptilian ancestors, amphibians are not the intermediate step in the evolution of reptiles from fishes.

Modern amphibians are united by several unique traits. They typically have a moist skin and rely heavily on cutaneous (skin-surface) respiration. They possess a double-channeled hearing system, green rods in their retinas to discriminate hues, and pedicellate (two-part) teeth. Some of these traits may have also existed in extinct groups.

Members of the three extant orders differ markedly in their structural appearance. Frogs and toads are tailless and somewhat squat with long, powerful hind limbs modified for leaping. In contrast, caecilians are limbless, wormlike, and highly adapted for a burrowing existence. Salamanders and newts have tails and two pairs of limbs of roughly the same size; however, they are somewhat less specialized in body form than the other two orders.

Many amphibians are obligate breeders in standing water. Eggs are laid in water, and the developing larvae are essentially free-living embryos; they must find their own food, escape predators, and perform other life functions while they continue to develop. As the larvae complete their embryonic development, they adopt an adult body plan that allows them to leave aquatic habitats for terrestrial ones. Even though this metamorphosis from aquatic to terrestrial life occurs in members of all three amphibian groups, there are many variants, and some taxa bear their young alive. Indeed, the roughly 6,200 living species of amphibians display more evolutionary experiments in reproductive mode than any other vertebrate group. Some taxa have aquatic eggs and larvae, whereas others embed their eggs in the skin on the back of the female; these eggs hatch as tadpoles or miniature frogs. In other groups, the young develop within the oviduct, with the embryos feeding on the wall of the oviduct. In some species, eggs develop within the female's stomach.

General Features

Size Range and Diversity of Structure

The three living orders of amphibians vary greatly in size and structure. The presence of a long tail and two pairs of limbs of about equal size distinguishes newts and salamanders (order Caudata) from other amphibians, although members of the eel-like family Sirenidae have no hind limbs. Newts and salamanders vary greatly in length; members of the Mexican genus Thorius measure 25 to 30 mm (1 to 1.2 inches), whereas Andrias, a genus of giant aquatic salamanders endemic to China and Japan, reaches a length of more than 1.5 metres (5 feet). Frogs and toads (order Anura) are easily identified by their long hind limbs and the absence of a tail. They have only five to nine presacral vertebrae. The West African goliath frog, which can reach 30 cm (12 inches) from snout to vent and weigh up to 3.3 kg (7.3 pounds), is the largest anuran. Some of the smallest anurans include the South American brachycephalids, which have an adult snout-to-vent length of only 9.8 mm (0.4 inch), and some microhylids, which grow to 9 to 12 mm (0.4 to 0.5 inch) as adults. The long, slender, limbless caecilians (order Gymnophiona) are animals that have adapted to fossorial (burrowing) lifestyles by evolving a body segmented by annular grooves and a short, blunt tail. Caecilians can grow to more than 1 metre (3 feet) long. The largest species, Caecilia thompsoni, reaches a length of 1.5 metres (5 feet), whereas the smallest species, Idiocranium russeli, is only 90 to 114 mm (3.5 to 5 inches) long.

Olympic torrent salamander (Rhyacotriton olympicus).

Distribution and Abundance

Amphibians occur widely throughout the world, even edging north of the Arctic circle in Eurasia; they are absent only in Antarctica, most remote oceanic islands, and extremely xeric (dry) deserts. Frogs and toads show the greatest diversity in humid tropical environments. Salamanders primarily inhabit the Northern Hemisphere and are most abundant in cool, moist, montane forests; however, members of the family Plethodontidae, the lungless salamanders, are diverse in the humid tropical montane forests of Mexico, Central America, and northwestern South America. Caecilians are found spottily throughout the African, American, and Asian wet tropics.

For many years, habitat destruction has had a severe impact on the distribution and abundance of numerous amphibian species. Since the 1980s, a severe decline in the populations of many frog species has been observed. Although acid rain, global warming, and ozone depletion are contributing factors to these reductions, a full explanation of the disappearance in diverse environment remains uncertain. A parasitic fungus, the so-called amphibian chytrid (Batrachochytrium dendrobatidis), however, appears to be a major cause of substantial frog die-offs in parts of Australia and southern Central America and milder events in North America and Europe.

Economic Importance

Amphibians, especially anurans, are economically useful in reducing the number of insects that destroy crops or transmit diseases. Frogs are exploited as food, both for local consumption and commercially for export, with thousands of tons of frog legs harvested annually. The skin secretions of various tropical anurans are known to have hallucinogenic effects and effects on the central nervous and respiratory systems in humans. Some secretions have been found to contain magainin, a substance that provides a natural antibiotic effect. Other skin secretions, especially toxins, have potential use as anesthetics and painkillers. Biochemists are currently investigating these substances for medicinal use.

Amazon leaf frog An Amazon leaf frog (Cruziohyla craspedopus) ascending a liana in an Ecuadorean rainforest. Amazon leaf frogs are found in lowland tropical forested areas of Colombia, Ecuador, Peru, and Brazil.

Reproduction

The three living groups of amphibians have distinct evolutionary lineages and exhibit a diverse range of life histories. The breeding behaviour of each group is outlined below. One similar tendency among amphibians has been the evolution of direct development, in which the aquatic egg and free-swimming larval stages are eliminated. Development occurs fully within the egg capsule, and juveniles hatch as miniatures of the adult body form. Most species of lungless salamanders (family Plethodontidae), the largest salamander family, some caecilians, and many species of anurans have direct development. In addition, numerous caecilians and a few species of anurans and salamanders give birth to live young (viviparity).

Anurans display a wide variety of life histories. Centrolenids and phyllomedusine hylids deposit eggs on vegetation above streams or ponds; upon hatching, the tadpoles (anuran larvae) drop into the water where they continue to develop throughout their larval stage. Some species from the families Leptodactylidae and Rhacophoridae create foam nests for their eggs in aquatic, terrestrial, or arboreal habitats; after hatching, tadpoles of these families usually develop in water. Dendrobatids and other anurans deposit their eggs on land and transport them to water. Female hylid marsupial frogs are so called because they carry their eggs in a pouch on their backs. A few species lack a pouch and the tadpoles are exposed on the back; in some species, the female deposits her tadpoles in a pond as soon as they emerge.

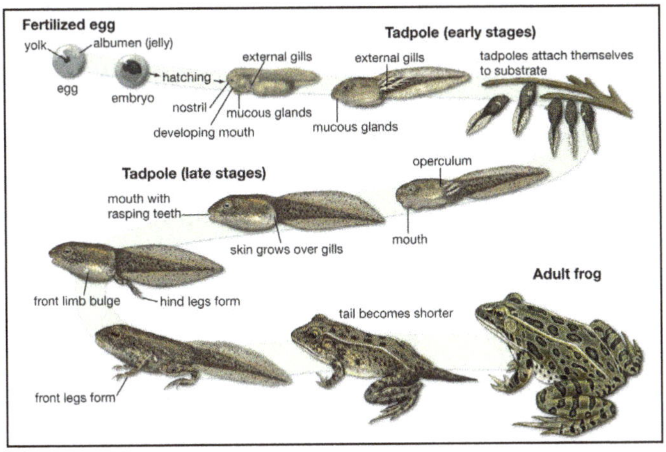

Life cycle of the European common frog (Rana temporaria).

Embryonic Stage

Inside the egg, the embryo is enclosed in a series of semipermeable gelatinous capsules and suspended in perivitelline fluid, a fluid that also surrounds the yolk. The hatching larvae dissolve these capsules with enzymes secreted from glands on the tips of their snouts. The original yolk mass of the egg provides all nutrients necessary for development; however, various developmental stages utilize different nutrients. In early development, fats are the major energy source. During gastrulation, an early developmental stage in which the embryo consists of two cell layers, there is an increasing reliance on carbohydrates. After gastrulation, a return to fat utilization occurs. During the later developmental stages, when morphological structures form, proteins are the primary energy source. By the neurula stage, an embryonic stage in which nervous tissue develops, cilia appear on the embryo, and the graceful movement of these hairlike structures rotates the embryo within the perivitelline fluid. The larvae of direct developing and live-bearing caecilians, salamanders, and some anurans have external gills that press against the inner wall of the egg capsule, which permits an exchange of gases (oxygen and carbon dioxide) with the outside air or with maternal tissues. During development, ammonia is the principal form of nitrogenous waste, and it is diluted by a constant diffusion of water in the perivitelline fluid.

The development of limbs in the embryos of aquatic salamanders begins in the head region and proceeds in a wave down the body, and digits appear sequentially on both sets of limbs. Salamanders that deposit their eggs in streams produce embryos that develop both sets of limbs before they hatch, but salamanders that deposit their eggs in still water have embryos that develop only forelimbs before hatching. (In contrast, the limbs of anurans do not appear until after hatching.) Soon after the appearance of forelimbs, most pond-dwelling salamanders develop an ectodermal projection known as a balancer on each side of the head. These rodlike structures arise from the mandibular arch, contain nerves and capillaries, and produce a sticky secretion. They keep newly hatched larvae from sinking into the sediment and aid the salamander in maintaining its balance before its forelimbs develop. After the forelimbs appear, the balancers degenerate.

During the embryonic and early larval stages in anurans, paired adhesive organs arise from the hyoid arch, located at the base of the tongue. The sticky mucus they secrete can form a threadlike attachment between a newly hatched tadpole and the egg capsule or vegetation. Consequently, the tadpole that is still developing can remain in a stable position until it is capable of swimming and feeding on its own, after which the adhesive organs degenerate.

Larval Stage

The amphibian larva represents a morphologically distinct stage between the embryo and adult. The larva is a free-living embryo. It must find food, avoid predators, and participate in all other aspects of free-living existence while it completes its embryonic development and growth. Salamander and caecilian larvae are carnivorous, and they have a morphology more like their respective adult forms than do anuran larvae. Not long after emerging from their egg capsules, larval salamanders, which have four fully developed limbs, start to feed on small aquatic invertebrates. The salamander larvae are smaller versions of adults, although they differ from their adult counterparts by the presence of external gills, a tailfin, distinctive larval dentition, a rudimentary tongue, and the absence of eyelids. Larval caecilians, also smaller models of adults, have external gills, a lateral-line system (a group of epidermal sense organs located over the head and along the side of the body), and a thin skin.

Principal types of salamander larvae.

In anurans, tadpoles are fishlike when they hatch. They have short, generally ovoid bodies and long, laterally compressed tails that are composed of a central axis of musculature with dorsal and ventral fins. The mouth is located terminally (recessed), ringed with an oral disk that is often fringed by papillae and bears many rows of denticles made of keratin. Tadpoles often have horny beaks. Their gills are internal and covered by an operculum. Water taken in through the mouth passes over the gills and is expelled through one or more spiracular openings on the side of an opercular chamber. Anuran larvae are microphagous and thus feed largely on bacteria and algae that coat aquatic plants and debris.

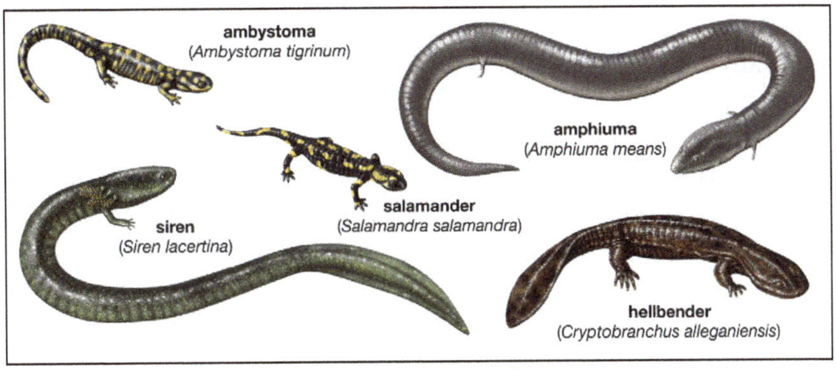

Representative urodeles.

Salamander larvae usually reach full size within two to four months, although they may remain larvae for two to three years before metamorphosis occurs. Some large aquatic species, such as the hellbender (Cryptobranchus alleganiensis) and the mud puppy (Necturus maculosus), never fully metamorphose and retain larval characteristics as adults. Tadpole development varies in length between species. Some anuran species living in xeric (dry) habitats, in which ephemeral ponds may exist for only a few weeks, develop and metamorphose within two to three weeks; however, most species require at least two months. Species living in cold mountain streams or lakes often require much more time. For example, the development of the tailed frog (Ascaphus truei) takes three years to complete.

Metamorphosis

Metamorphosis entails an abrupt and thorough change in an animal's physiology and biochemistry, with concomitant structural and behavioral modifications. These changes mark the transformation from embryo to juvenile and the completion of development. Hormones ultimately control all events of larval growth and metamorphosis, and in many instances, development is accompanied by a shift from a fully aquatic life to a semiaquatic or fully terrestrial one. Although salamanders undergo many structural modifications, these changes are not dramatic. The skin thickens as dermal glands develop and the caudal fin is resorbed. Gills are resorbed and gill slits close as lungs develop and branchial (gill) circulation is modified. Eyelids, tongue, and a maxillary bone are formed, and teeth develop on the maxillary and parasphenoid bones.

Changes that occur in caecilians—the closure of the gill slit, the degeneration of the caudal fin, and the development of a tentacle and skin glands—are also minor.

Skeletal changes are much more dramatic in anurans because tadpoles make an abrupt and radical transition to their adult form. Limbs complete their development, and the forelimbs break through the opercular wall, early in metamorphosis. The tail shrinks as it is resorbed by the body, dermal glands develop, and the skin becomes thicker. As lungs and pulmonary ventilation develop, gills and their associated blood circulation disappear. Adult mouthparts replace their degenerating larval equivalents, and hyolaryngeal structures develop. All anurans except pipids (family Pipidae) develop a tongue. In the newly differentiated digestive tract, the intestine is shortened. The eyes become larger and are structurally altered; eyelids appear. These extreme changes of anuran metamorphosis clearly demarcate the shift from an aquatic to a terrestrial mode of life. Other less obvious yet nonetheless radical modifications of the larval skull and hyobranchial apparatus (that is, the part of the skeleton that serves as base for the tongue on the floor of the mouth) occur to make room for newly developed sense organs. These modifications also facilitate the transition from larval modes of feeding and respiration to those of the adult.

During metamorphosis, the urogenital system of all amphibians is also modified. A mesonephric or opisthonephric kidney—which uses nephrons located either in the middle or at the end of the nephric ridge in the developing embryo—replaces the degenerating rudimentary pronephric kidney. This transition is linked to the shift from production of a large volume of dilute ammonia to a small amount of concentrated urea. Gonads and associated ducts also appear and begin their maturation.

Heterochrony

Neoteny, once a widely used label for the condition of sexually mature larvae, has been discontinued by biologists and replaced by the concept of heterochrony. Heterochrony refers to the change in the timing and rate of developmental events, and it is a widespread feature in amphibian evolution, particularly in salamanders. During development, a structure can begin to develop sooner (predisplacement) or later (postdisplacement) in an organism than it occurred in the ancestral species or parents. Also, a structure may continue to develop beyond the previous embryological sequence (hypermorphosis) or the developmental sequence can stop earlier than normal (progenesis or hypomorphosis). Each of these heterochronic events can produce a structurally or functionally different organism.

The classical "neotenic" salamander, the axolotl (Ambystoma mexicanum), is a paedomorphic species (that is, a species that retains aspects of its juvenile form during its adult phase); it retains its larval gills. In the mole salamander (Ambystoma talpoideum), some populations also display hypomorphic development in which the development of several larval traits to the adult condition is delayed. Since the gonads mature, a population of sexually mature salamanders with a larval morphology is produced.

Heterochrony also explains the presence of larval traits in adults of the salamander families Cryptobranchidae and Proteidae (olms and mud puppies).

Heterochrony is not confined to salamanders. The different sized eardrums in the American bullfrog (Lithobates catesbeianus) are examples of hypermorphism in male bullfrogs. The development of the eardrums in the male extends beyond that of the female.

Bullfrog (Lithobates catesbeianus).

Life Cycle

Many amphibians have a biphasic life cycle involving aquatic eggs and larvae that metamorphose into terrestrial or semiaquatic juveniles and adults. Commonly, they deposit large numbers of eggs in water; clutches of the tiger salamander (Ambystoma tigrinum) may exceed 5,000 eggs, and large bullfrogs (L. catesbeianus) may produce clutches of 45,000 eggs. Egg size and water temperature are important factors that influence an embryo's development time. Eggs of many anuran species laid in warm water require only one or two days to develop, whereas eggs deposited in cold mountain lakes or streams may not hatch for 30 to 40 days. The development of salamander eggs often requires more time, with hatching occurring 20 to 270 days after fertilization.

Food and Feeding

Adult amphibians consume a wide variety of foods. Earthworms are the main diet of burrowing caecilians, whereas anurans and salamanders feed primarily on insects and other arthropods. Large salamanders and some large anurans eat small vertebrates, including birds and mammals. Most anurans and salamanders locate prey by sight, although some use their sense of smell. The majority of salamanders and diurnal (that is, active during daylight) terrestrial anurans are active foragers, but many other anurans employ a sit-and-wait technique. Caecilians locate their underground prey with a chemosensory tentacle and capture their quarry with a powerful bite. Aquatic salamanders lunge at their prey with an open mouth and appear to suck the victim in by expanding their buccal (oral) cavity. The terrestrial lunged salamander extends its sticky tongue, which is attached anteriorly to the floor of the mouth, to ensnare a meal. In lungless salamanders, the hyobranchial apparatus is not part of the process of buccal respiration; this apparatus is modified so that it can project the tongue from the mouth. The end of the tongue is sticky to adhere to prey, and prey can be captured at distances ranging from 40 to 80 percent of the salamander's body length.

Primitive anurans have feeding mechanisms that resemble those of the typical terrestrial salamanders. More advanced anurans employ a "lingual flip," in which the surfaces of the retracted tongue are twisted and inverted in the fully extended tongue. The pipids, which are completely aquatic, are unique among anurans; they lack a tongue and thus must essentially suck food and water into their mouth.

Form and Function

Although the structure of the muscular, skeletal, and other anatomical systems are specifically modified for each group, amphibians are often set apart from other groups of animals by their characteristic skin, or integument, and evolutionary advances in vision and hearing.

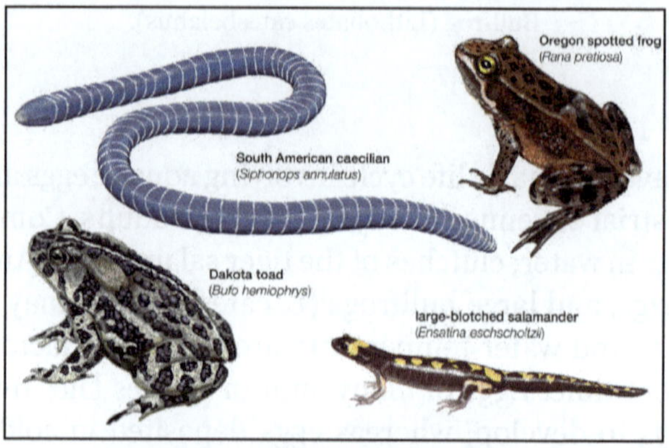

Representative amphibians.

The circulatory and respiratory systems work with the integument to provide cutaneous respiration. A broad network of cutaneous capillaries facilitates gas exchange and the diffusion of water and ions between the animal and the environment. Several species of salamanders and at least one species of frog (Barbourula kalimantanensis) are lungless. Amphibians also employ various combinations of branchial and pulmonary strategies to breathe. The buccal pump mechanism, which involves the pushing of air between the lungs and the closed mouth, is present in amphibians and some groups of fishes.

In addition to its roles in respiration and maintaining water balance, the integument of amphibians contains poison glands that release toxins. Specific toxins are found only in amphibians and are used to defend against predators.

The eye of the modern amphibian (or lissamphibian) has a lid, associated glands, and ducts. It also has muscles that allow its accommodation within or on top of the head, depth perception, and true colour vision. These adaptations are regarded as the first evolutionary improvements in vertebrate terrestrial vision. Green rods in the retina, which permit the perception of a wide range of wavelengths, are found only in lissamphibians.

The amphibian auditory system is also specially adapted. One modification is the papilla amphibiorum, a patch of sensory tissues that is sensitive to low-frequency sound. Also unique to lissamphibians is the columella-opercular complex, a pair of elements associated with the auditory capsule that transmit airborne (columella) or seismic (operculum) signals.

Frog

Frog is the common name for any of the members of the amphibian order Anura, whose extant species are characterized by an adult with longer hind legs among the four legs, a short body, webbed digits, protruding eyes, and the absence of a tail. Anurans have well-developed voices, whereas the other two orders of amphibians are limited to sounds such as coughs and grunts.

Frogs are the most numerous and diverse amphibians, being found in nearly all habitats, including arboreal, aquatic, and terrestrial niches, and every continent except Antarctica. Three species have ranges that extend above the Arctic Circle. The greatest diversity is in tropical rainforests. Overall, about 88 percent of amphibian species are frogs, with the order Anura containing 5,250 species in 33 families, of which the Leptodactylidae, Hylidae and Ranidae are the richest in species.

A distinction is often made between frogs and toads on the basis of their appearance, with toad the common term imprecisely applied to largely terrestrial members of Anura that are characterized by short legs, a stocky body, and a drier, warty or bumpy skin and frogs those members that are aquatic or semi-aquatic with slender bodies, longer legs, and smooth and/or moist skins.

However, this division of anurans into toads and frogs is a popular, not a scientific distinction; it does not represent a formal taxonomic rank. From a taxonomic perspective, all members of the order Anura are frogs. The only family exclusively given the common name "toad" is Bufonidae, the "true toads," although many species from other families are also called toads. The anuran family "Ranidae" is known as the "true frogs."

Most anurans have a semi-aquatic lifestyle, but move easily on land by jumping or climbing. They typically lay their eggs in puddles, ponds, or lakes, and their larvae, called tadpoles, have gills and develop in water. Though adults of some species eat plants, adult frogs of almost all species follow a carnivorous diet, mostly of arthropods, annelids, and gastropods. Some tadpoles are carnivorous as well. Frogs are most noticeable by their call, which can be widely heard during the night or day, mainly in their mating season.

Frogs provide many ecological, commercial, scientific, and cultural values. Ecologically, they are integral to many aquatic and terrestrial food chains. Commercially, they are raised as a food source, and scientifically and educationally, they have served as an important model organism throughout the history of science and today dead frogs are used for dissections in anatomy classes. Culturally, frogs feature prominently in folklore, fairy tales, and popular culture. In addition, the unique morphology and behavior of frogs, including their calls and life cycle, add greatly to the wonder of nature for humans.

Although they are among the most diverse groups of vertebrates, populations of certain frog species are significantly declining.

Morphology and Physiology

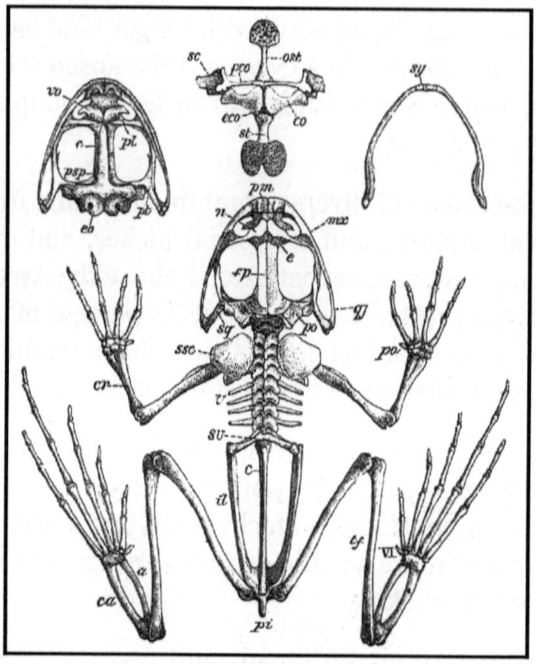

Skeleton of Rana

The morphology of frogs is unique among amphibians. Compared with the other two groups of amphibians (salamanders and caecilians), frogs are unusual because they lack tails as adults and their legs are more suited to jumping than walking.

The physiology of frogs is generally like that of other amphibians (and differs from other terrestrial vertebrates) because oxygen can pass through their highly permeable skin. This unique feature allows frogs to "breathe" largely through their skin. Because the oxygen is dissolved in an aqueous film on the skin and passes from there to the blood, the skin must remain moist at all times; this makes frogs susceptible to many toxins in the environment, some of which can similarly dissolve in the layer of water and be passed into their bloodstream. This may be a cause of the decline in frog populations.

Despite the common name of toad, the European fire-bellied toad (Bombina bombina) prefers a watery habitat

Many characteristics are not shared by all of the approximately 5,250 described frog species. However, some general characteristics distinguish them from other amphibians. Frogs are usually well-suited to jumping, with long hind legs and elongated ankle bones. They have a short vertebral column, with no more than ten free vertebrae, followed by a fused tailbone (urostyle or coccyx), typically resulting in a tailless phenotype.

Frogs range in size from 10 millimeters (Brachycephalus didactylus of Brazil and Eleutherodactylus iberia of Cuba) to 300 millimeters (goliath frog, Conraua goliath, of Cameroon). The skin hangs loosely on the body because of the lack of loose connective tissue. Skin texture varies: it can be smooth, warty, or folded.

In the head area, frogs have three eyelid membranes: one is transparent to protect the eyes underwater, and two vary from translucent to opaque. Frogs have a tympanum on each side of the head, which is involved in hearing and, in some species, is covered by skin. Most frogs do, in fact, have teeth of a sort. They have a ridge of very small cone teeth around the upper edge of the jaw. These are called maxillary teeth. Frogs often also have what are called vomerine teeth on the roof of their mouth. They do not have anything that could be called teeth on their lower jaw, so they usually swallow their food whole. The so-called "teeth" are mainly used to hold the prey and keep it in place till they can get a good grip on it and squash their eyeballs down to swallow their meal. True toads, however, do not have any teeth.

Feet and Legs

The structure of the feet and legs varies greatly among frog species, depending in part on whether they live primarily on the ground, in water, in trees, or in burrows. Frogs must be able to move quickly through their environment to catch prey and escape predators, and numerous adaptations help them do so.

Tyler's Tree Frog (Litoria tyleri) illustrates large toe pads and webbed feet.

Many frogs, especially those that live in water, have webbed toes. The degree to which the toes are webbed is directly proportional to the amount of time the species lives in the water. For example, the completely aquatic African dwarf frog (Hymenochirus sp.) has fully webbed toes, whereas the toes of White's tree frog (Litoria caerulea), an arboreal species, are only a half or a quarter webbed.

Arboreal frogs have "toe pads" to help grip vertical surfaces. These pads, located on the ends of the toes, do not work by suction. Rather, the surface of the pad consists of interlocking cells, with a small gap between adjacent cells. When the frog applies pressure to the toe pads, the interlocking cells grip irregularities on the substrate. The small gaps between the cells drain away all but a thin layer of moisture on the pad, and maintain a grip through capillarity. This allows the frog to grip smooth surfaces, and does not function when the pads are excessively wet.

In many arboreal frogs, a small "intercalary structure" in each toe increases the surface area touching the substrate. Furthermore, since hopping through trees can be dangerous, many arboreal frogs have hip joints that allow both hopping and walking. Some frogs that live high in trees even possess an elaborate degree of webbing between their toes, as do aquatic frogs. In these arboreal frogs, the webs allow the frogs to "parachute" or control their glide from one position in the canopy to another.

Ground-dwelling frogs generally lack the adaptations of aquatic and arboreal frogs. Most have smaller toe pads, if any, and little webbing. Some burrowing frogs have a toe extension—a metatarsal tubercle—that helps them to burrow. The hind legs of ground dwellers are more muscular than those of aqueous and tree-dwelling frogs.

Skin

Many frogs are able to absorb water directly through the skin, especially around the pelvic area. However, the permeability of a frog's skin can also result in water loss. Some tree frogs reduce water loss with a waterproof layer of skin. Others have adapted behaviors to conserve water, including engaging in nocturnal activity and resting in a water-conserving position. This position involves the frog lying with its toes and fingers tucked under its body and chin, respectively, with no gap between the body and substrate. Some frog species will also rest in large groups, touching the skin of the neighboring frog. This reduces the amount of skin exposed to the air or a dry surface, and thus reduces water loss. These adaptations only reduce water loss enough for a predominantly arboreal existence, and are not suitable for arid conditions.

Common Eastern Froglet (Crinia signifera) camouflaged against leaf litter.

Camouflage is a common defensive mechanism in frogs. Most camouflaged frogs are nocturnal, which adds to their ability to hide. Nocturnal frogs usually find the ideal camouflaged position during the day to sleep. Some frogs have the ability to change color, but this is usually restricted to shades of one or two colors. For example, White's tree frog varies in shades of green and brown. Features such as warts and skin folds are usually found on ground-dwelling frogs, where a smooth skin would not disguise them effectively. Arboreal frogs usually have smooth skin, enabling them to disguise themselves as leaves.

Certain frogs change color between night and day, as light and moisture stimulate the pigment cells and cause them to expand or contract.

Poison

Oophaga pumilio, a poison dart frog, contains numerous alkaloids which deter predators.

Many frogs contain mild toxins that make them distasteful to potential predators. For example, all toads have large poison glands—the parotid glands—located behind the eyes on the top of the head. Some frogs, such as some poison dart frogs, are especially toxic. The chemical makeup of toxins in frogs varies from irritants to hallucinogens, convulsants, nerve poisons, and vasoconstrictors. Many predators of frogs have adapted to tolerate high levels of these poisons. Others, including humans, may be severely affected.

Some frogs obtain poisons from the ants and other arthropods they eat; others, such as the Australian Corroboree frogs (Pseudophryne corroboree and Pseudophryne pengilleyi), can manufacture an alkaloid not derived from their diet.

Some native people of South America extract poison from the poison dart frogs and apply it to their darts for hunting, although few species are toxic enough to be used for this purpose. It was previously a misconception the poison was placed on arrows rather than darts. The common name of these frogs was thus changed from "poison arrow frog" to "poison dart frog" in the early 1980s.

Poisonous frogs tend to advertise their toxicity with bright colors, an adaptive strategy known as aposematism. There are at least two non-poisonous species of frogs in tropical America (Eleutherodactylus gaigei and Lithodytes lineatus) that mimic the coloration of dart poison frogs' coloration for self-protection (Batesian mimicry).

Because frog toxins are extraordinarily diverse, they have raised the interest of biochemists as a "natural pharmacy." The alkaloid epibatidine, a painkiller 200 times more potent than morphine, is found in some species of poison dart frogs. Other chemicals isolated from the skin of frogs may offer resistance to HIV infection. Arrow and dart poisons are under active investigation for their potential as therapeutic drugs.

The skin secretions of some toads, such as the Colorado River toad and cane toad, contain bufotoxins, some of which, such as bufotenin, are psychoactive, and have therefore been used as recreational drugs. Typically, the skin secretions are dried and smoked. Skin licking is especially dangerous, and appears to constitute an urban myth.

Respiration and Circulation

The skin of a frog is permeable to oxygen and carbon dioxide, as well as to water. There are a number of blood vessels near the surface of the skin. When a frog is underwater, oxygen is transmitted through the skin directly into the bloodstream. On land, adult frogs use their lungs to breathe. Their lungs are similar to those of humans, but the chest muscles are not involved in respiration, and there are no ribs or diaphragm to support breathing. Frogs breathe by taking air in through the nostrils (causing the throat to puff out), and compressing the floor of the mouth, which forces the air into the lungs.

Frogs are known for their three-chambered heart, which they share with all tetrapods except birds and mammals. In the three-chambered heart, oxygenated blood from the lungs and de-oxygenated blood from the respiring tissues enter by separate atria, and are directed via a spiral valve to the appropriate vessel—aorta for oxygenated blood and pulmonary vein for deoxygenated blood. This special structure is essential to keeping the mixing of the two types of blood to a minimum, which enables frogs to have higher metabolic rates, and to be more active than otherwise.

The life cycle of frogs, like that of other amphibians, consists of the main stages of egg, tadpole, metamorphosis, and adult. The reliance of frogs on an aquatic environment for the egg and tadpole stages gives rise to a variety of breeding behaviors that include the well-known mating calls used by the males of most species to attract females to the bodies of water that they have chosen for breeding. Some frogs also look after their eggs—and in some cases even the tadpoles—for some time after laying.

Life Cycle

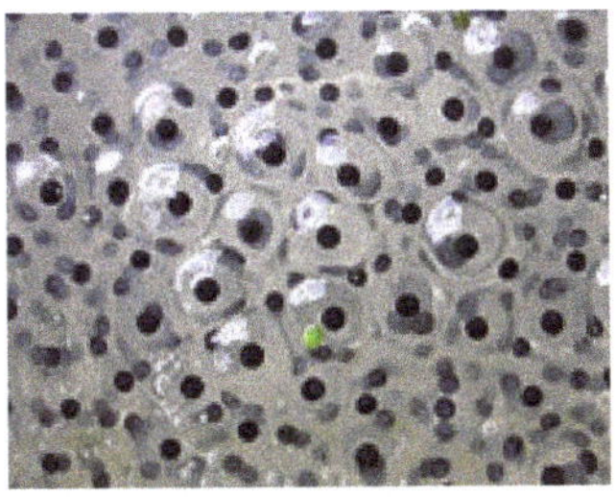

Frogspawn

Tadpoles

Adult leopard frog

The life cycle of a frog starts with an egg. A female generally lays frogspawn, or egg masses containing thousands of eggs, in water. While the length of the egg stage depends on the species and environmental conditions, aquatic eggs generally hatch within one week.

Eastern Banjo Frog in water

The eggs are highly vulnerable to predation, so frogs have evolved many techniques to ensure the survival of the next generation. Most commonly, this involves synchronous reproduction. Many individuals will breed at the same time, overwhelming the actions of predators; the majority of the offspring will still die due to predation, but there is a greater chance some will survive. Another way in which some species avoid the predators and pathogens eggs are exposed to in ponds is to lay eggs on leaves above the pond, with a gelatinous coating designed to retain moisture. In these species, the tadpoles drop into the water upon hatching. The eggs of some species laid out of water can detect vibrations of nearby predatory wasps or snakes, and will hatch early to avoid being eaten. Some species, such as the cane toad (Bufo marinus), lay poisonous eggs to minimize predation.

Eggs hatch and the frogs continue life as tadpoles (occasionally known as polliwogs). Tadpoles are aquatic, lack front and hind legs, and have gills for respiration and tails with fins for swimming. Tadpoles are typically herbivorous, feeding mostly on algae, including diatoms filtered from the water through the gills. Some species are carnivorous at the tadpole stage, eating insects, smaller tadpoles, and fish. Tadpoles are highly vulnerable to predation by fish, newts, predatory diving beetles, and birds such as kingfishers. Cannibalism has been observed among tadpoles. Poisonous tadpoles are present in many species, such as cane toads. The tadpole stage may be as short as a week, or tadpoles may overwinter and metamorphose the following year in some species, such as the midwife toad (Alytes obstetricans) and the common spadefoot (Pelobates fuscus).

At the end of the tadpole stage, frogs undergo metamorphosis, in which they transition into adult form. Metamorphosis involves a dramatic transformation of morphology and physiology, as tadpoles develop hind legs, then front legs, lose their gills, and develop lungs. Their intestines shorten as they shift from an herbivorous to a carnivorous diet. Eyes migrate rostrally and dorsally, allowing for binocular vision exhibited by the adult frog. This shift in eye position mirrors the shift from prey to predator, as the tadpole develops and depends less upon a larger and wider field of vision and more upon depth perception. The final stage of development from froglet to adult frog involves apoptosis (programmed cell death) and resorption of the tail.

After metamorphosis, young adults may leave the water and disperse into terrestrial habitats, or continue to live in the aquatic habitat as adults. Almost all species of frogs are carnivorous as adults, eating invertebrates such as arthropods, annelids, and gastropods. A few of the larger species may eat prey such as small mammals, fish, and smaller frogs. Some frogs use their sticky tongues to catch fast-moving prey, while others capture their prey and force it into their mouths with their hands. There are a very few species of frogs that primarily eat plants. Adult frogs are themselves preyed upon by birds, large fish, snakes, otters, foxes, badgers, coatis, and other animals. Frogs are also eaten by people.

Reproduction of Frogs

Once adult frogs reach maturity, they will assemble at a water source such as a pond or stream to breed. Many frogs return to the bodies of water where they were born, often resulting in annual migrations involving thousands of frogs. In continental Europe, a large proportion of migrating frogs used to die on roads, before special fences and tunnels were built for them.

Male and female Tyler's tree frogs in amplexus (male is typical breeding yellow coloration.)

Once at the breeding ground, male frogs call to attract a mate, collectively becoming a chorus of frogs. The call is unique to the species, and will attract females of that species. Some species have satellite males who do not call, but intercept females that are approaching a calling male.

The male and female frogs then undergo amplexus. This involves the male mounting the female and gripping her tightly. Fertilization is external: the egg and sperm meet outside of the body. The female releases her eggs, which the male frog covers with a sperm solution. The eggs then swell and develop a protective coating. The eggs are typically brown or black, with a clear, gelatin-like covering.

Most temperate species of frogs reproduce between late autumn and early spring. In the United Kingdom, most common frog populations produce frogspawn in February, although there is wide variation in timing. Water temperatures at this time of year are relatively low, typically between four and 10 degrees Celsius. Reproducing in these conditions helps the developing tadpoles because dissolved oxygen concentrations in the water are highest at cold temperatures. More importantly, reproducing early in the season ensures that appropriate food is available to the developing frogs at the right time.

Parental Care

Although care of offspring is poorly understood in frogs, it is estimated that up to 20 percent of amphibian species may care for their young in one way or another, and there is a great diversity of parental behaviors. Some species of poison dart frogs lay eggs on the forest floor and protect them, guarding the eggs from predation and keeping them moist. The frog will urinate on them if they become too dry. After hatching, a parent (the sex depends upon the species) will move them, on its back, to a water-holding bromeliad. The parent then feeds them by laying unfertilized eggs in the bromeliad until the young have metamorphosed.

Color plate from Ernst Haeckel's 1904 Kunstformen der Natur, depicting frog species that include two examples of parental care.

Other frogs carry the eggs and tadpoles on their hind legs or back (e.g., the midwife toads). Some frogs even protect their offspring inside their own bodies. The male Australian pouched frog (Assa darlingtoni) has pouches along its side in which the tadpoles reside until metamorphosis. The female gastric-brooding frogs (genus Rheobatrachus) from Australia, now probably extinct, swallows its tadpoles, which then develop in the stomach. To do this, the gastric-brooding frog must stop secreting stomach acid and suppress peristalsis (contractions of the stomach). Darwin's frog (Rhinoderma darwinii) from Chile puts the tadpoles in its vocal sac for development. Some species of frog will leave a "babysitter" to watch over the frogspawn until it hatches.

Call

The call of a frog is unique to its species. Frogs call by passing air through the larynx in the throat. In most calling frogs, the sound is amplified by one or more vocal sacs, membranes of skin under the throat or on the corner of the mouth that distend during the amplification of the call. Some frog calls are so loud that they can be heard up to a mile away.

Some frogs lack vocal sacs, such as those from the genera Heleioporus and Neobatrachus, but these species can still produce a loud call. Their buccal cavity is enlarged and dome-shaped, acting as a resonance chamber that amplifies their call. Species of frog without vocal sacs and that do not have a loud call tend to inhabit areas close to flowing water. The noise of flowing water overpowers any call, so they must communicate by other means.

The main reason for calling is to allow males to attract a mate. Males call either individually or in a group called a chorus. Females of many frog species, for example Polypedates leucomystax, produce calls reciprocal to the males', which act as the catalyst for the enhancement of reproductive activity in a breeding colony. A male frog emits a release call when mounted by another male. Tropical species also have a rain call that they make on the basis of humidity cues prior to a rain shower. Many species also have a territorial call that is used to chase away other males. All of these calls are emitted with the mouth of the frog closed.

A distress call, emitted by some frogs when they are in danger, is produced with the mouth open, resulting in a higher-pitched call. The effectiveness of the call is unknown; however, it is suspected the call intrigues the predator until another animal is attracted, distracting them enough for its escape.

Salamander

Salamanders are characterized by slender bodies, short legs, and long tails. They are most closely related to the caecilians, little-known legless amphibians. Most of the animals in the salamander order look like a cross between a lizard and a frog. They have moist, smooth skin like frogs and long tails like lizards.

Salamanders are found in most moist or arid habitats in the Northern Hemisphere, but can also be found south of the equator. They live on all continents except Antarctica and Australia. Salamanders live in or near water or on moist ground, often in a swamp. Some species live in water most of their life, some live their entire adult life on land, and some live in both habitats. Some salamanders live in caves. These salamanders have pale skin and reduced eyes as they have adapted to living in complete darkness in underground pools of water. The reduced eyes are similar to other organisms that live in caves or underground.

Salamanders are carnivorous, eating only other animals, not plants. They will eat almost any smaller animal, such as worms, centipedes, crickets, spiders, and slugs. Some will even eat small invertebrates. Finally, salamanders have the ability to grow back lost limbs, as well as other body parts. This process is known as regeneration.

Salamanders have developed ways not to be eaten. Most salamanders have brightly colored, poisonous skin. The bold color tells predators not to eat the salamander. Many salamanders have glands on the back of the neck or on the tail that give off a poisonous or bad-tasting liquid. Some species can even shed their tail during an attack and grow a new one later. Some salamanders stand high on its legs and waves its tail to scare away danger. One particular salamander, the ribbed newt, has needle-like rib tips. It can squeeze its muscles to make the rib tips pierce through its skin and into its enemy, telling the predator to stay away, a feature unique among the animal kingdom.

The marbled salamander (left) shows the typical salamander body plan: slender body, short legs, long tail, and moist skin. Caecilian (right) are a type of legless amphibian most closely related to salamanders.

Breathing Process in Salamanders

Different salamanders breathe in different ways. In those that have gills, breathing occurs through the gills as water passes over the gill slits. Sirens keep their gills all their lives, which allows them to breathe underwater.

Species that live on land lose their gills as they grow older. These salamanders develop lungs that are used in breathing, much like breathing in mammals. Other land-living salamanders do not have lungs or gills. These are called lungless salamanders. Instead, they "breathe," or exchange gases, through their skin. This requires blood vessels that exchange gases to be spread throughout the skin.

Size of Salamanders

Salamanders are generally small. However, some can reach a foot or more, as in the mudpuppy of North America. In Japan and China, the giant salamander reaches 6 feet and weighs up to 66 pounds.

The Pacific giant salamander can reach up to 6 feet in length and weigh up to 66 pounds. Classification of Salamanders.

Classification of Salamanders

Salamanders belong to a group of approximately 500 species of amphibians. The order Urodela, containing salamanders and newts, is divided into three suborders:

1. Giant salamanders, including the hellbender and Asiatic salamanders.
2. Advanced salamanders, including lungless salamanders, mudpuppies, and newts. Newts are salamanders that spend most of each year living on land.
3. Sirens are salamanders that have lungs as well as gills and never develop beyond the larval stage.

Sirens have only two legs, but the other salamander species develop four legs as adults, with fleshy toes at the end of each foot. The legs on four-legged salamanders are so short that the salamander belly drags on the ground as the animal walks. Sirens have long, strong tails that are flat to help sirens swim like a fish, with the tail swinging from side to side.

Distribution

The Blue-spotted Salamander occurs in eastern central North America in a broad band stretching from the Atlantic provinces and northern New England, around the Great Lakes, and west as far as central Manitoba.

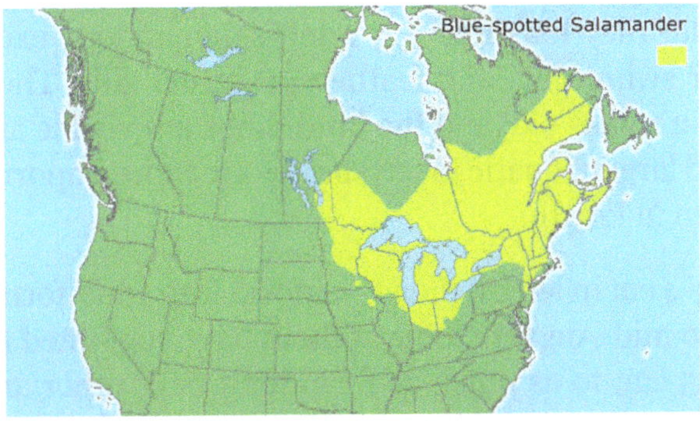

In Manitoba, the species is found in the southeastern corner of our province, east of the Red River and as far north as the narrows of Lake Winnipeg. However, this is a very cryptic species, and one for which intensive inventory hasn't occurred. There will likely be extensions of its known range in this province for some time into the future.

Habitat

This is a woodland species that is probably restricted to areas of moist forest floor, near lakes, streams and wetlands in areas with mixed forests (deciduous and coniferous trees together). It breeds in ephemeral, snow-melt ponds, like those used by wood frogs, and in small permanent ponds that are free of fish.

Populations

There are no estimates available of the population densities of the Blue-spotted Salamander. It is not a widely studied species, and its cryptic nature makes it very hard to gather data. Blue-spotted Salamanders should be the focus of more research in this province.

Life Cycle

Blue-spotted Salamanders breed in early spring, from April to mid-May, in small ponds. The adults emerge from winter hibernation and gather in the ponds. Most of the mating activity likely occurs at night. For most salamanders, the males use a combination of pheromones and visual displays to attract females into close proximity. The Blue-spotted has a prolonged courtship where the male holds onto the female with his front legs and rubs his chin over her head. The male deposits a packet of sperm (a spermatophore) on the pond bottom, and tries to draw the female over top of it. She then picks it up with her cloaca (urogenital opening) and fertilizes her eggs internally.

Females lay their eggs, as many as 500, individually or in small bunches at the base of sticks, plants or rocks in the pond. The eggs take about a month to hatch, depending on water temperature. When they hatch the larvae have well developed eyes, mouths, external gills and a broad tail fin, so they can see, lunge and snap at the tiny aquatic animals they feed on, such as daphnia, copepods or other invertebrates.

As the larvae get older they look like tiny adult salamanders, except they have broad fins on their tails for swimming, and feathery, external gills. The front limbs emerge when they are about 2 weeks old, and the hind limbs are visible after about 3 weeks of age. Insect larvae and larger aquatic invertebrates are more important in their diet once they reach about 2 cm in length.

They grow to about 3-4 cm in length then transform into adult form and leave the pond, usually by late July to mid-August. Some authors have suggested that the larvae transform when they are 5 cm, to as much as 8 cm long. Like other amphibians, the exact length of time required to reach a size where they can transform, and even the ultimate size they do transform at, will vary with food availability and other conditions. They lose their external gills and tail fins, and gain their adult colouration before leaving the pond. It's thought they require about two years to reach adult size. How long they live for is anybody's guess. Other Ambystomid salamanders can live for more than 10 years.

References

- Herpetology: newworldencyclopedia.org, Retrieved 23 July, 2019
- Reptile, animal: britannica.com, Retrieved 3 February, 2019
- Snake: newworldencyclopedia.org, Retrieved 28 April, 2019

- Structure-of-snakes-discussed-reptiles, reptiles, zoology: biologydiscussion.com, Retrieved 8 June, 2019
- McDiarmid, Roy W. (2012). "Reptile Diveristy and Natural History: An Overview". In McDiarmid, Roy W.; et al. (eds.). Reptile Biodiversity: Standard Methods for Inventory and Monitoring. p. 13. ISBN 978-0520266711
- Crocodile-order, animal: britannica.com, Retrieved 11 May, 2019
- Batrachology-assignment: myassignmenthelp.net, Retrieved 17 July, 2019
- Amphibian, animal: britannica.com, Retrieved 9 January, 2019
- Salamanders-MS-LS, salamanders, biology: ck12.org, Retrieved 6 March, 2019
- Bluespot, creature, spring: naturenorth.com, Retrieved 15 August, 2019

Permissions

All chapters in this book are published with permission under the Creative Commons Attribution Share Alike License or equivalent. Every chapter published in this book has been scrutinized by our experts. Their significance has been extensively debated. The topics covered herein carry significant information for a comprehensive understanding. They may even be implemented as practical applications or may be referred to as a beginning point for further studies.

We would like to thank the editorial team for lending their expertise to make the book truly unique. They have played a crucial role in the development of this book. Without their invaluable contributions this book wouldn't have been possible. They have made vital efforts to compile up to date information on the varied aspects of this subject to make this book a valuable addition to the collection of many professionals and students.

This book was conceptualized with the vision of imparting up-to-date and integrated information in this field. To ensure the same, a matchless editorial board was set up. Every individual on the board went through rigorous rounds of assessment to prove their worth. After which they invested a large part of their time researching and compiling the most relevant data for our readers.

The editorial board has been involved in producing this book since its inception. They have spent rigorous hours researching and exploring the diverse topics which have resulted in the successful publishing of this book. They have passed on their knowledge of decades through this book. To expedite this challenging task, the publisher supported the team at every step. A small team of assistant editors was also appointed to further simplify the editing procedure and attain best results for the readers.

Apart from the editorial board, the designing team has also invested a significant amount of their time in understanding the subject and creating the most relevant covers. They scrutinized every image to scout for the most suitable representation of the subject and create an appropriate cover for the book.

The publishing team has been an ardent support to the editorial, designing and production team. Their endless efforts to recruit the best for this project, has resulted in the accomplishment of this book. They are a veteran in the field of academics and their pool of knowledge is as vast as their experience in printing. Their expertise and guidance has proved useful at every step. Their uncompromising quality standards have made this book an exceptional effort. Their encouragement from time to time has been an inspiration for everyone.

The publisher and the editorial board hope that this book will prove to be a valuable piece of knowledge for students, practitioners and scholars across the globe.

Index

CPSIA information can be obtained
at www.ICGtesting.com
Printed in the USA
LVHW060905030122
707669LV00002B/365

9 781641 727068